Next-Gen Manufacturing 2018

Topical Conference at the 2018 AIChE Annual Meeting

Pittsburgh, Pennsylvania, USA
28 October - 2 November 2018

ISBN: 978-1-5108-7637-8

Printed from e-media with permission by:

Curran Associates, Inc.
57 Morehouse Lane
Red Hook, NY 12571

Some format issues inherent in the e-media version may also appear in this print version.

Printed by Curran Associates, Inc. (2019)

For permission requests, please contact AIChE
at the address below.

AIChE
120 Wall Street, FL 23
New York, NY 10005-4020

Phone: (800) 242-4363
Fax: (203) 775-5177

www.aiche.org

Additional copies of this publication are available from:

Curran Associates, Inc.
57 Morehouse Lane
Red Hook, NY 12571 USA
Phone: 845-758-0400
Fax: 845-758-2633
Email: curran@proceedings.com
Web: www.proceedings.com

TABLE OF CONTENTS

Some abstracts in this book refer to images that were not made available by the publisher

535550 Development of Nanoparticle Alignment Regimes in Drying Cellulose Nanocrystal Droplet Suspensions for Additive Manufacturing

Abstract:

Nanoparticle suspensions have potential for a number of drop-on-demand applications, where they are deposited in droplet form and subsequently dried via evaporation of the suspending agent. The droplet drying process often results in deposition of ring patterns, typically referred to as the coffee ring effect. When drying picoliter-sized, aqueous, cellulose nanocrystal droplet suspensions, which are on the size scale of traditional drop-on-demand technologies, we observe alignment of these rod-like nanoparticles in multiple orientations. A radial alignment is formed near the center of the droplet, and tangential orientation is observed in the particles along the circumference of the droplet. Tuning particle orientation enables unique opportunities for control of nanocomposite structure required for next generation optical and electronic materials that can be generated using drop-on-demand or voxel based technologies. To successfully control the nanoparticle orientation during the drying process, a fundamental understanding of the alignment mechanism is necessary.

We have developed a 3-dimensional, continuum-based, finite element model implementing a deforming mesh to understand the impact of shear induced alignment driven by evaporative flow. Simulations of a single-phase, sessile water droplet suggest a relationship between velocity profile, shear stress, and particle orientation. We have explored the coupled effects of Marangoni flows, air convection, heat transport, and concentration dependent viscosity models on evaporation of single-phase picoliter-sized droplets leading to new insights into particle orientation in the evaporation of colloidal droplets that result from a complex interplay of initial concentration, substrate surface functionality, and particle-particle interactions. Velocity and shear stress profiles indicate the development of radial flows enacting shear stresses on the fluid driving the alignment of rod-like nanoparticles in the flow direction which varies with location in the confined flow field.

522259 Characterization of Poly(ether imide) Towards the Development of a Fused Filament Fabrication (FFF) Process Model

Abstract:

Fused filament fabrication (FFF) is the most common method of AM that relies on thermoplastic extrusion. As the AM field is transitioning from prototype visualization to functional prototypes and end product development, one of the greatest hurdles is the inherent z-axis (interlayer) weakness of the processed parts resulting from weak adhesion between deposited layers, producing anisotropic mechanical properties. This poor adhesion stems from insufficient polymer inter-diffusion between the layers, referred to as â€œself-healing.â€

To better understand the evolution of part strength during the FFF process, we are developing a model to describe the self-healing phenomenon that couples the complex thermal and geometric profiles with diffusion at the layer interface. This model is strongly dependent on the transient behavior of the polymer governed by the processing and environmental conditions present during the printing process. Additionally, we are examining the effect of adding a continuous composite fiber with anisotropic thermal conductivity. Our preliminary work has developed a realistic three-dimensional model of a printed part which will accurately represent the interface between individual layers. Previous models have utilized idealistic models based on circles and ovals of equal size. Our results prove the individual layers are flattened ellipsoids that vary in diameter depending on their height from the heated bed. We have also analyzed the complete re-entanglement time of the neat polymer to quantify the extent of self-healing, combined with characterizing the thermal conductivity of the composite filaments. The corresponding mechanical strength is driven by interlayer adhesion and characterized as a function of processing conditions. Our preliminary rheological findings of a poly(ether imide) amorphous thermoplastic indicate that times for complete re-entanglement range from 47 seconds to more than 22 minutes for isothermal temperatures of 375 °C and 345 °C, respectively. Since the cooling process is non-isothermal, and the layers remain above their glass transition temperatures for less than 3 seconds, this would limit the amount of chain re-entanglement at the layer interface and corresponding bond strength between layers of a printed part, and suggests that a modified printing process is required to create stronger parts with this material. The inclusion of the composite fiber does increase the time at which the temperature of the polymer is elevated and should help increase the amount of self-healing that occurs during the printing process.

537447 Additive Manufacturing of Core-Shell Microparticles Containing Thermosetting Resins

Abstract:

Cold spray deposition is a common additive manufacturing method for fabricating metal coatings that involves the spraying metal particles at high-speeds. However, this scalable, additive manufacturing process has not yet generated solid polymer coatings. Our study enables the cold spraying of solid polymer materials at room temperature. Here, we developed novel, core-shell thermoplastic microparticles containing a high concentration of thermosetting resins. Suspension polymerization was employed to produce the microparticles due to its many advantages, namely low conversion cost, mature scale-up processing, and a versatile tunability in terms of particle size and morphology. The thermal properties, size distribution, morphology, and composition of the particles were explored and fine-tuned in an effort to optimize the deposition of the particles at room temperature using cold spray technology. The outcome of this study is a novel form of sprayable thermosetting powders that deposit to form a coating with a high performance, easy operation processing, significant environmental benefits, and a commercial feasibility.

532737 Supersonic-Impaction Printing of Flame-Made Doped-Perovskite Nanoparticles

Abstract:

Perovskite nanoparticles find a range of applications in emerging electronics such as transparent conductors, solar cells, light-emitting diodes and sensors such as gas sensing. Such electroactive oxides typically have melting temperatures over 1000°C. Subsequently, their deposition, calcination and sintering also require extremely high temperatures. As a result, substrate compatibility with high temperature treatments is needed to facilitate usability and implementation of such materials. This is therefore a bottleneck for their application on substrates that are susceptible to high temperatures and oxidizing environments such as plastics or metals. More recently, the advent of additive manufacturing has shed new light on the deposition and growth of oxide materials. Among them, perovskite materials synthesized via sol-gel methods followed by inkjet and aerosol printing has shown promise for being able to be deposited on low temperature substrates. In this presentation, we will be showcasing a new concept and design for a deposition chamber intended for the supersonic impaction of aerosolized nanomaterials. Further, we will also discuss the implementation of a flame spray reactor as the source of synthesizing pure Barium Stannate ($BaSnO_3$) nanoparticles that are impacted using supersonic deposition to create electrically conducting perovskite thin films.

The principle of supersonic deposition is primarily based on acceleration of an aerosol-containing gas stream through a converging-diverging (de Laval) nozzle. This de Laval nozzle creates a high pressure drop across itself when downstream is pumped to moderate vacuum and upstream is maintained at a higher pressure. The pressure gradient across the nozzle accelerates the flowing aerosol to supersonic speeds and subsequently impact them onto the substrate. The kinetic energy of the particles is converted into heat upon impaction and thus the process can create a homogeneous thin film of sintered materials. Through computational fluid dynamics simulations, we found a number of parameters including the effect of gas velocities, nanoparticle density, size, trajectories and other physical properties to affect the efficiency of impaction. Unlike other precursors for competing additive manufacturing techniques such as inkjet, aerosol and electro-hydrodynamic jet printing, supersonic impaction does not require the use of any surfactant stabilized nanoparticle inks.

For this reason, flame spray pyrolysis was chosen as a scalable route for the synthesis of pure aerosolized ceramic nanoparticles without the use of any surfactants. More recently, flames are also being implemented for the synthesis of complex oxides such as tertiaries and perovskites. We found that the critical parameters that affect the flame temperature are the precursor and dispersion oxygen flow rates and enthalpy of the precursor material. A higher feed rate and precursors with high enthalpies ensure higher temperatures of the flame. Doped $BaSnO_3$ is an electroactive perovskite material and has a large indirect intrinsic band-gap of 3.2 eV. We will discuss our developments pertaining to the incorporation of electrical properties in the thin films of $BaSnO_3$ *via* vacuum annealing to induce oxygen vacancies and incorporation of external dopants that can substitute the Ba^{2+} site (n-type with La) or Sn^{4+} sites (p-type with K, Cs).

541761 Simulating Powder Handling Processes in Additive Manufacturing Using the Discrete Element Method

Abstract:

Additive Manufacturing is a rapidly advancing manufacturing area that enables users to produce complex parts in a short amount of time. One common approach to AM in industry involves the deposition and selective sintering/melting of successive layers of metallic powders. Through repeated build-up of powder layers, complex metal parts can be produced.

There are a wide range of metal powders available for AM production and each has its own characteristics. During printing a powder must be stored, transported, and processed within the printing equipment to produce the final part. Each of these stages need to be performed efficiently by the printer to ensure an optimum printing process.

The specific composition and behavioral traits of the powder used in AM is highly influential on printer set-up and resulting part manufacturing quality. Introducing a new powder into an AM process means that the set-up of the printing device needs to be configured to handle this new material effectively. Because of the variability of powder properties, flowability and susceptibility to environmental conditions, it is not straight-forward to make decisions as to how to reach the optimal configuration. At present, a common method to adjust printer set-up to account for the influence of a specific powder requires trial and error testing, or a reliance on know-how from previous work with the same printer and powder combinations. This takes time and extends the overall part production cycle.

Discrete Element Modelling (DEM) is a simulation method for recreating bulk and granular material behaviors in a virtual environment. It is a method used in other powder-focused industries, such as pharmaceutical, to provide engineering insight into powder handling and processing operations. By using DEM simulation, engineers can test design and process concepts and extract meaningful information from virtual versions of their material and equipment to help in the decision-making process.

The same opportunities exist in AM, and such a simulation method offers engineers an alternative to physical trial and error testing when trying to understand how key stages in part production will perform. In this presentation we will discuss the application of DEM, using EDEM software, to the additive manufacturing print process. Examples will be given to demonstrate how DEM can be used to help engineers understand the performance of powder delivery systems, as well as gaining insight into metrics such as powder porosity variation during the powder-layering process.

533694 3D Printed Complex Dosage Forms Embedded with Engineered BCS Class II Drug Particles

Abstract:

Recent work suggests complex structured dosage forms can be manufactured where drug particles can be integrated via 3D printing technology. However, incorporating engineered drug particles into 3D printed structures so that various intermediate and final product properties can be enhanced, for example, good mechanical properties and drug content uniformity of feed materials, has yet to be explored. It is hypothesized that engineered particles enable uniform distribution of drug into molten polymeric matrix via twin screw extruder (TSE). This distinct feature of engineered particles leads to obtaining enhanced dissolution profile from 3D structured dosage forms. The printable filaments containing varying drug loading (DL) %, and different extent of particle engineering as well as filament their mechanical properties were considered in this investigation. Griseofulvin (GF) and Hydroxypropyl cellulose (HPC) SL grade was used as the model BCS Class II drug and the polymeric matrix, respectively. The feed materials having four different DL% (5, 10, 15, and 25 wt.% GF) incorporating engineered and non-engineered particles were extruded by HME and 3D structured tablets from the feed materials were successfully created via FDM 3D printer. No drug degradation was observed in TGA results for any cases. All the drug particles kept their crystalline structure from HME extrusion (140 °C) as well as FDM printing (170 °C). The feed materials containing engineered particles exhibited higher tensile strength and better modulus of elasticity, regardless of the drug loading. Moreover, 3D printed tablets containing engineered particles showed enhanced drug release for all DL%. Overall, the 3D printed tablets embedded with engineered drug particles resulted in tablets without any degradation/recrystallization issues and demonstrated enhanced dissolution profiles for four different dosage amounts.

529744 3D Printing for Rapid Prototyping of Innovative Process Equipment for Pharmaceutical Crystallization

Abstract:

The development of pharmaceuticals involves various stages with preclinical and clinical trials. A short time-to-market is essential to maximize the patent protection and to avoid any loss in potential revenues.[1] Furthermore, most pharmaceuticals in development do not reach commercial manufacturing. However, for those pharmaceuticals that do reach market, efficient and novel processing equipment can provide a competitive edge. Hence, rapid prototyping of processing equipment during development is important to minimize time-to-market, while maximizing early learning on processing. 3D printing is an attractive technology to fabricate customized and novel processing equipment almost instantaneously. However, despite this potential, the use of 3D printing during early pharmaceutical process development is still in its infancy.

3D printing has found applications for the fabrication of micro-reactors and to some extent for mesoscale reactors as well.[2] In contrast to reactors, the use of 3D printing to fabricate innovative designs of separators has received limited attention. For example, for crystallization, which is a key unit operation in many pharmaceutical processes, no studies on the use of 3D printing for rapid prototyping have been reported to the best of our knowledge. Pharmaceutical crystallization is complicated by the need to optimize intrinsic crystal quality attributes such as size distribution, which affects the performance of downstream processes and possibly the efficacy of the final product. Optimization of crystallization requires control over various mixing processes. 3D printing can be a flexible and rapid method to fabricate innovative structures of crystallizers for optimized mixing at all relevant scales during pharmaceutical development.

The objective of this work is to develop and characterize 3D printed prototypes of innovative crystallization equipment with different structures for mixing. Two cases will be investigated. A common feature of both cases is that the mixing mechanism does not require moving parts, which makes fabrication with 3D printing attractive.

The first case involves a 3D-printed airlift crystallizer for protein crystallization. Protein crystallization in an airlift crystallizer has not been reported yet, but is attractive as protein crystals are generally more fragile in comparison to crystals of small-molecule active ingredients.[3] Therefore, mild and homogeneous mixing conditions are required, which can be provided by airlift columns due to the low maximum shear forces.[4] An airlift crystallizer was designed and fabricated with a desktop stereolithography (SLA) printer. The hydrodynamics of the column have been characterized for lysozyme as model protein. A sufficiently high upward velocity in the column demonstrates practical feasibility for protein crystallization in an airlift crystallizer. Next, the crystallization performance of the airlift crystallizer is characterized in this work and compared to a conventional stirred tank crystallizer.

The second case involves antisolvent and reactive crystallization in continuous flow. Rapid mixing is achieved by using static mixers. The effectiveness of kenics-type of static mixers has previously been demonstrated for anti-solvent crystallization of ketoconazole [5]. In this work, Y-mixers and komax-type mixers have been designed and fabricated using a desktop stereolithography (SLA) 3D printer. Preliminary studies on reactive crystallization of barium sulfate have been conducted, which demonstrate practical feasible and reduced agglomeration when using the komax mixer in comparison to when using a Y-mixer. The stability of many stereolithography resins used for 3D-printed structures is good for aqueous-based system. However, pharmaceutical crystallization often involves organic compounds. Therefore, this case particularly focuses on the use of various 3D printing techniques to address stability issues for organic solvents. In the long term, this research aims to develop a platform around 3D printing, which allows users to select, design and fabricate innovative crystallization equipment rapidly to achieve shorter time-to-market and intensified pharmaceutical crystallization.

Acknowledgement: The work described in this abstract was supported by a grant from the Research Grants Council of the Hong Kong Special Administrative Region, China (Project No. 16242916).

References

[1] G. Reklaitis, J. Khinast, and F. J. Muzzio, "Pharmaceutical engineering science—New approaches to pharmaceutical development and manufacturing," Chem. Eng. Sci., vol. 65, no. 21, pp. iv–vii, Nov. 2010.

[2] A. K. Au, W. Huynh, L. F. Horowitz, and A. Folch, "3D-Printed Microfluidics," Angew. Chemie - Int. Ed., vol. 55, no. 12, pp. 3862–3881, 2016.

[3] S. Tait, E. T. White, and J. D. Litster, "Mechanical Characterization of Protein Crystals", Part. Part. Syst. Charact., vol. 25, no. 3, pp. 266-276, Sep 2008.

[4] A. Soare, R. Lakerveld, J. Van Royen, G. Zocchi, A. I. Stankiewicz, and H. J. M. Kramer, "Minimization of attrition and breakage in an airlift crystallizer," Ind. Eng. Chem. Res., vol. 51, no. 33, pp. 10895–10909, Aug 2012.

[5] A. J. Alvarez and A. S. Myerson, "Continuous plug flow crystallization of pharmaceutical compounds," Cryst. Growth Des., vol. 10, no. 5, pp. 2219–2228, May 2010.

529577 Direct Write of UV Curable Polymer Bonded Magnets

Abstract:

Today, almost all commercial magnet producers are using the conventional sintering-based procedure to manufacture magnets. Polymer-bonded hard magnets have recently gained popularity as a novel magnet manufacturing technique due to its design freedom regarding shapes and magnetizing structures. Comparing to the conventional sintering process, the maximum energy product $(BH)_{max}$ of the polymer-bonded magnets is lower. However, the fabrication of polymer-bonded magnets does not require high temperature treatment and thus can be more efficient. Furthermore, no current mass production technologies can produce magnets with complex geometries.

This work utilizes extrusion-based direct write (DW), combined with UV curing, to fabricate polymer-bonded magnets at room temperature. DW is an additive manufacturing technique enabling the deposition of electronic components and functional or structural patterns using different types of materials, directly without utilizing masks or subsequent etching processes. Compared to the molding process, DW provides intricate details and smooth surface finish. In this work, Nd-Fe-B powder are directly mixed with photo-curable binders to form a printable DW ink. Experiments were carried out to understand the relationship between magnetic powder solid loading, processing, and mechanical strength. Magnetic performance of printed parts was compared to other polymer-bonded and sintered magnets. The study opens a new pathway for preparing polymer-bonded magnets at room temperature for various magnetic applications.

523345 Before You Click "Print": Regulatory Considerations for 3D Printed Oral Drug Products

Abstract:

Additive manufacturing/3D printing of oral drug products and medical devices has become more attractive to shift the manufacturing strategies to support individualized medicines of the future. The US Food and Drug Administration (FDA) is charged with protecting and promoting the public health by ensuring these products are safe and effective. To that end, FDA's recent approval of a 3D-printed drug product in August 2015 (SPRITAM®) introduced a new chapter in pharmaceutical manufacturing for solid oral dosage form. Although the first approved 3D-printed drug product is a solid oral immediate release product, the majority of 3D printing research for oral delivery has been focused on controlled release, targeting and precise delivery for extremely low dose drugs. Various principles of 3D printing of oral drug products could support production of complex and personalized medicines on demand. However, standardization of some aspects of the 3D printed dosage form and associated manufacturing process may be required to achieve this objective. In this presentation, technical and regulatory considerations for development and control of 3D printed oral products will be discussed for various 3D printing technologies.

DISCLAIMER

This presentation reflects the views of the presenter and should not be construed to represent FDA's views or policies.

531710 Two-Color Photo-Inhibited Systems for Rapid Additive Manufacturing

Abstract:

Additive manufacturing (AM), also known as 3D-printing, is exciting disruptive technology that may upend contemporary manufacturing practices. Production in AM generally proceeds by layer-by-layer addition of materialâ €"typically thermoplastic or photopolymerizable resinâ€"to gradually build up the desired geometry, allowing for the production of highly individualized parts with near limitless complexity. The achievable print speed is still one of the major limitations of the current technologies with layer-by-layer devices only able to achieve vertical print speeds of several millimeters per hour. Recently, a layer-less (or continuous) method for resin based AM, termed continuous liquid interface production, was developed. The method relies on a thin diffusion-controlled oxygen-inhibited dead zone to eliminate adhesion to the projection window during printing, allowing for an order of magnitude increase in the achievable printing speeds when compared with traditional SLA devices.

Photo-inhibitor molecules that inhibit polymerization in the presence of light may be an attractive alternative inhibition mechanism to oxygen-inhibited systems. The advantages of this type of system are that diffusion of oxygen no longer controls the thickness of the dead zone, and the limit of this zone is theoretically much larger. Larger dead zones allow faster polymer reflow under the printing part and, therefore, faster build speeds. This talk will present a novel method using a two-color irradiation scheme to spatially control polymerization to generate dead zones for rapid AM and enable additional functionality over conventional continuous additive manufacturing methods.

531942 Glass-Forming Polymer Networks for Shape-Memory Contact Printing

Abstract:

Current methods for microscale pattern transfer fail to combine high feature resolution, ability to pattern large areas, high throughput and cost effectiveness. Over the past years, several researchers have demonstrated how films of shape-memory polymers can exhibit topographical memory, leading to switchable adhesion or optical properties. We are designing and developing new stimuli-responsive materials to ultimately enable high fidelity pattern transfer by thermal modulation of the elastomeric contact printing process.

Different glass forming polymers that can withstand high elongation and store large amounts of elastic strain energy in compression are under investigation. Two systems will be described: (i) small molecule thiol-ene networks that are softened by addition of dithiol reagents, and (ii) glass forming t-butyl acrylates that are lightly crosslinked. Their shape-memory behavior in tension and in compression will be reported. Furthermore, a custom-built materials testing / contact printing apparatus will be used to set compressive strain between two soft surfaces while measuring stress and contact area. Contact mechanics experiments are modeled and interpreted using finite element analysis to offer deeper insight to the nature of contact and detachment.

540924 Nano- and Microfabricated Hydrogels for Regenerative Engineering

Abstract:

Engineered materials that integrate advances in polymer chemistry, nanotechnology, and biological sciences have the potential to create powerful medical therapies. Our group aims to engineer tissue regenerative therapies using water-containing polymer networks, called hydrogels, that can regulate cell behavior. Specifically, we have developed photocrosslinkable hybrid hydrogels that combine natural biomolecules with nanoparticles to regulate the chemical, biological, mechanical and electrical properties of gels. These functional scaffolds induce the differentiation of stem cells to desired cell types and direct the formation of vascularized heart or bone tissues. Since tissue function is highly dependent on architecture, we have also used microfabrication methods, such as microfluidics, photolithography, bioprinting, and molding, to regulate the architecture of these materials. We have employed these strategies to generate miniaturized tissues. To create tissue complexity, we have also developed directed assembly techniques to compile small tissue modules into larger constructs. It is anticipated that such approaches will lead to the development of next-generation regenerative therapeutics and biomedical devices.

Commercial Scale Manufacturing of a Pharmaceutical Product Using Powder-Liquid 3D Printing Technology

Timothy S. Tracy, PhD, Aprecia Pharmaceuticals, Blue Ash, OH

Abstract

Three dimensional printing (3DP) opens the possibility to manufacture pharmaceuticals differently from today and provides the opportunity to create novel dosage forms for patients. Despite significant advances in the field, an obstacle to commercialization of 3DP pharmaceuticals has been the difficulty in being able to manufacture at commercial scale. Recently, utilizing powder-liquid 3DP technology a pharmaceutical (levetiracetam, Spritam™) has been developed through the 505(b)(2) regulatory pathway at commercial quantities and has received FDA approval for manufacture and sale. In addition, the product is a novel dosage form that utilizes fast-melt technology (ZipDose™) that provides for rapid dispersion in the mouth (< 10 seconds) with only a sip of liquid, while containing doses of levetiracetam up to 1000mg. Products that rapidly disperse in the mouth are particularly useful for patients with dysphagia or difficulty swallowing, particularly of large tablets. This rapid dispersion at high doses was possible because of the selective porosity and hardness of the dosage form created by the 3DP layering process. To produce this unique dosage form, processes were developed to assure consistent and precise jetting of print fluid and careful control of powder deposition across up to 30 layers at commercial production speed to produce tablets that were highly accurate and reproducible with respect to dose of active drug. The regulatory approval and commercialization of this product demonstrated that unique pharmaceutical dosage forms could be created through 3DP technology that helped address unmet needs.

540925 Closing the Circle on Design, Hardware, and Materials for Additive Manufacturing

Abstract:

3D printing is a subset of additive manufacturing (AM) and represents an evolving set of tools and capabilities to prototype, and in some cases produce, new products in much faster iteration cycles than traditional subtractive processes. For 3D printing to reach its full potential as a viable large-scale manufacturing option, new advancements are still required across the three major segments of the industry â€“ design, hardware, and materials â€“ and to accelerate these advancements, highly collaborative efforts between the segments will provide 3D printing users with the quickest and best solutions for their production needs. The presentation will highlight the current state-of-the-art in materials for 3D printing, known existing property gaps, Dowâ€™s vision for the future development and performance of 3D printable materials, and how the broader industry can collaborate to shorten the path to wider scale adoption of these tools in 21st century manufacturing.

541541 Additive Manufacturing and Architected Materials

Abstract:

Material properties are governed by the chemical composition and spatial arrangement of constituent elements at multiple length-scales. This fundamentally limits material properties with respect to each other creating trade-offs when selecting materials for specific applications. For example, strength and density are inherently linked so that, in general, the more dense the material, the stronger it is in bulk form. We are combining advanced microstructural design, using flexure and screw theory as well as topology optimization, with advanced additive micro and nanomanufacturing techniques to create new material systems with previously unachievable property combinations â€" mechanical metamaterials. The performance of these materials is fundamentally controlled by geometry at multiple length-scales rather than chemical composition. We have demonstrated designer properties of these mechanical metamaterials in polymers, metals, ceramics and combinations thereof. Properties include ultra-stiff lightweight materials, negative stiffness, and negative thermal expansion. Our additive micro-manufacturing techniques include Projection Microstereolithography (PmSL), Direct Ink Writing (DIW), and Electrophoretic Deposition (EPD). These tools are capable of generating the designed structures which are highly three-dimensional micro- and nano-scale architectures with multiple constituent materials in the same structure.

530486 Cellulose Nanocrystal Thermoplastic Urethane Composite Filament for Fused Filament Fabrication

Abstract:

Improving material selection for the fused filament fabrication (FFF) process has gained significant attention in recent years. Several researchers and companies have been pushed by consumers and industry to stretch the capabilities of the FFF process. FFF was originally developed as a prototyping tool and is now progressively challenged each day to produce end use parts for a wide variety of applications. We have built on the development of a smart mechanically dynamic material, one which can respond to changes in environment with minimal energy input, to potentially increase design freedom and realize new opportunities for expansion of FFF into new functional products.

This study focuses on the production of thermoplastic urethane (TPU)/ cellulose nanocrystal (CNC) filaments for use in FFF. We investigate the influence of various melt processing parameters on the stimuli-responsiveness of the adaptive thermoplastic nanocomposite filament. The composite system analyzed is composed of Texin® RxT70A with 10 wt% dispersed CNC, generated through a mast batching process. Randomly oriented film samples are created via solvent casting and uniaxially oriented samples are created via single screw melt extrusion. The orientation of each sample is characterized by both polarized Raman spectroscopy and small angle X-ray scattering (SAXS). Dynamic mechanical analysis results indicate that uniaxial orientation does not significantly impact dry or wet state modulus, likely resulting from a change in the percolating nature of the CNCs. Thermal history imparted by the single screw extrusion process has a negligible impact on the functionality of the CNCs and the corresponding responsiveness of the composite.

532617 Direct Printing of Epoxy-Graphite Composite Ink for Thermal Management Devices

Abstract:

The project focuses on developing composite inks composed of a low viscosity Epoxy (EPON 862), graphite nanoplatelet and fillers such as carbon fibers, nanotubes and nanowires for use in thermal management devices. The viscoelastic properties are tuned by filler concentrations â€“ graphite and carbon fiber concentrations. By combining these functional materials 3D structures with a high thermal conductivity (~ 2 W/mK), high elastic modulus (> 1000MPa) and high electrical conductivity are achieved. Of particular focus is the relationship of rheological properties with final printed performance. The optimal graphite concentration in relation to the percolation threshold of graphite in the epoxy is determined and properties (thermal and electrical conductivity) measured as a function of graphite concentration. It is found out that the rheological percolation does not necessarily correspond to electrical percolation and a higher concentration is needed to achieve optimal properties. The results show that by carefully tailoring functional properties of the ink, net-shape multifunctional structures can be directly printed for thermal management device applications, such as heat sinks

532776 *Radio Frequency Heating of Carbon Nanotube Composite Materials for Additive Manufacturing*

Abstract:

Here, we report RF electromagnetic heating of polymer nanocomposite materials via direct contact and capacitively-coupled electric field applicators. Notably, RF heating allows nanocomposite materials to be resistively heated with applied electric fields. We highlight our novel RF heating technique for multi-walled carbon nanotube (MWCNT)/polymer composites and measure their broadband dielectric properties as a function of CNT loading. We also demonstrate two applications of this technique: (i) A key obstacle facing 3D-printed thermoplastic parts in engineering applications is the weak weld between successive filament traces, which often leads to delamination and mechanical failure. Here we report a novel concept for welding 3D-printed thermoplastic interfaces using intense localized heating of carbon nanotubes (CNTs) by microwave irradiation. We apply CNT-loaded coatings to 3D printer filament; after printing, microwave irradiation is shown to improve the weld fracture strength by 275%. (ii) We also demonstrate the use of RF heating to cure an automotive-grade epoxy loaded with MWCNTs. Our results show that lap shear joints cured faster with the RF method compared to control samples cured in an oven due to the heat transfer advantages of directly heating the epoxy composite.

536337 Rheology of Cement-Based Pastes for 3-D Printing Applications

Abstract:

Many industrial production processes are turning to additive manufacturing technologies as a promising alternative to reduce production costs and time, to add design flexibility and move towards more sustainable processing and improve manufacturing safety. The construction industry is a more recent sector to join the additive manufacturing revolution and researchers around the world are pushing the limit to demonstrate the printability of large-scale structural components. However, research on the fundamental materials science of printed cementitious bodies already lags far behind. To ensure sustained gains from additive manufacturing in construction, researchers must address issues regarding printability and microstructural continuity of printed infrastructure materials.

This work seeks to develop well-formulated additive systems aimed at enhancing cement-based pastes for effective printing. Various combinations of polymers including superabsorbent polymer, specialty polymers, viscosity modifying admixtures, superplasticizers and mineral-based additives, with desirable properties, and cement were explored. Various time-domain rheology experiments were performed to explore the flow properties, i.e. the yield stress and plastic viscosity, of select pastes were measured using oscillatory and flow rheometry. Pastes were also printed in different geometries to explore the process operating variables including extrusion speed, layer thickness and print rate. The research goal of this work is the identification of suitable mix formulations and printing conditions for effective deposition.

538763 Modeling of Cement Paste for 3-D Printing Applications

Abstract:

Additive manufacturing offers the construction industry new opportunities to reduce labor costs and waste, and increase the reliability of concrete infrastructure. The 3D printing of cement-based materials involves the, extrusion of cementitious mixtures through a nozzle to layer-by-layer build structural components. The aim of this study is to explore the relationships between printing conditions and paste physical properties using computational modeling. The key factor governing the properties of cement-based extrusion mixtures is the volumetric content and rheology of the cement paste, a non-Newtonian fluid with measurable yield stress "τ_0," and plastic viscosity "μ_0." COMSOL Multiphysics was used to analyze cement flows in 2-D and 3-D geometries relevant to printing of cement-based structures. The free surface flow of spherical, cylindrical, rectangular and elliptical gobs of cement paste in 2-D and 3-D wasstudied. A relationship between the fluid properties, printability metrics and scale (size of printed shapes) is suggested.

544816 CESMII Overview

Abstract:

The Clean Energy Smart Manufacturing Innovation Institute (CESMII), established by the US Department of Energy, is part of a network of institutes under the Manufacturing USA Initiative with a goal to secure the future of manufacturing in the U.S. through innovation, collaboration and education. CESMIIâ€™s mission is to radically democratize and accelerate the development and adoption of Smart Manufacturing (SM) technologies (sensing, control, data analytics, predictive modeling and platforms) to become the driving, sustainable engine that delivers energy efficiency and real-time business improvements in U.S. manufacturing. CESMII integrated strategies to achieve this mission include business practice innovation, enabling technologies, workforce development and a collaborative Smart Manufacturing Platformâ„¢ for industrial applications. Haresh Malkani, Chief Technology Officer â€“ CESMII, will provide an overview of the institute, its technology focus areas and recent progress.

544817 CESMII Roadmap Projects – Advancing CESMII's Mission

Abstract:

CESMII is about to kick off a portfolio of 10 innovative R&D roadmap projects covering a range of technologies including sensing, process control, data analytics, modeling, as well as workforce development, to address significant challenges faced by the manufacturing community. The projects cut across manufacturing types (discrete, hybrid, process), industry verticals (steel, cement, chemical, aerospace), and scale of focus (unit process, plant, enterprise). Miguel Corcio, Director - Technology Programs - CESMII, we will provide brief overview of the portfolio during this session.

546992 Panel Session: Democratization of Smart Manufacturing (SM)

Abstract:

Industrie 4.0, Smart Manufacturing, Industrial Internet of Things, Big Data, Cloud Computing are no longer just buzz words. They have become a reality and every year more manufacturers are getting started on this journey. But it takes investment in technology and talent to do so. Does it mean that this reality is only within the reach of the large and mighty companies? What can small and medium manufacturers do to be in this game? How do we make technology and expertise accessible for them cost effectively? Join our panelists representing both industry and academia in this healthy dialogue.

Panelists:

- Larry Megan – Praxair
- Sergio Butkewitsch-Choze – Arconic
- Panagiotis Christofides – UCLA
- B. Wayne Bequette – Rensselaer Polytechnic Institute

Moderator:

- Jim Davis, Vice Provost, Information Technology and Chief Academic Technology Officer, UCLA

530209 Development and Evaluation of Oxidative Dehydrogenation Technology for Production of Light Olefins

Abstract:

Light oleï¬ns are traditionally produced by the steam pyrolysis of saturated hydrocarbons as well as by recovery from refinery cracking operations. These technologies results in a wide range of byproducts, some valuable and some that need disposal. All of the conventional technology require extensive separation and purification.The technology known as RMD for reaction, membranes and distillation utilizies the M1 catalyst developed by Idaho National Laboratories and features very high selectivity at various levels of conversion in studies on the conversion of ethane to ethylene. Additionally, INL has found that the produced ethylene is refractory in additional reactors. This allows the interchanging of reactors and membranes, both of which provide high concentrations of ethylene. Process development and the associated economic analyses were performed using a cost of production procedure (COP) commonly called a required netback (RNB) analysis. These results were analyzed by incorporating the expected performance of a variety of $R_xM_yD_z$ arrangements with x,y, and z representing the number of the process segments. So for instance, $R_2M_1D_1$ indicates 2 ODH reactors, 1 facilitated transport (FTS) membrane and one heat pump distillation section. Results indicate that the combination of ODH reactors with FTS membranes is a new route that has much lower variable costs, ï¬xed costs and capital recovery that can reach shutdown economics level versus some existing technology.

530251 Process Intensification through Membrane Separation of Olefins and Paraffins

Abstract:

Ethylene and propylene are two of the major building blocks for the chemical industry. The separation of ethylene and propylene from their alkane counterparts is highly energy-intensive. It is estimated that separating alkanes from alkenes through distillation uses 0.3% of energy globally. In addition, distillation columns to separate these molecules are very capital-intensive. A membrane based olefin-paraffin separation technology has the potential to provide substantial economic benefit to the petrochemical industry and to drastically reduce the energy required for this separation. In addition, membrane systems are scalable, allowing their implementation for smaller, stranded streams where distillation columns have not previously been installed due to limited economic value. Membrane processes utilizing facilitated transport membranes have been extensively studied for separating olefin/paraffin mixtures. While separations have been demonstrated in the laboratory, problems with membrane stability have prevented development of commercial systems.

Compact Membrane Systems (CMS) has developed the Optiperm^{TM} membrane, a customized amorphous fluoropolymer (CAF) containing silver ions that selectively transport olefin molecules from a mixture of olefin and paraffin. The technology exploits a reversible complexation reaction mechanism known as facilitated transport. Silver incorporated into the membrane polymer backbone acts as a binding site for the double bond in the olefins transporting them across the membrane as depicted in the cartoon below. Saturated paraffins do not have the same interaction. This difference in affinity allows for olefins to pass through the membrane at a significantly higher rate than do paraffins.

This talk will share the latest field results from both the pilot currently running at the Delaware City Refining Company as well as any results coming out of the RAPID-funded collaboration with Dow Chemical. The DCRC pilot is operating on a stream with a 10-20% concentration of propylene while the Dow Chemical work will focus on higher olefins concentrations of both C2 and C3 streams. This will include performance data, longevity and membrane stability results as well as any learnings from these projects. Thus far, the DCRC pilot has demonstrated over 45 days of stable membrane operation with propylene permeance >75 GPU and selectivity >20 for propylene over propane.

535284 2D Optimization of Fixed-Bed Reactors: Additional Degrees of Freedom for the Reactor Design to Increase Efficiency

Abstract:

Highly efficient reactors are fundamental for resource- and energy efficient production of chemicals. To enable high reactor efficiencies, the reactor design aims to realize optimal reaction conditions along the reaction route [1]. In this regard, the reaction temperature plays a superior role due to its strong influence of the reaction rates.

For heterogeneously catalyzed reactions with a pronounced energy release, the state-of-the-art approach is the use of tube bundle reactors with axial dilution zones to approximate the optimal temperature [2]. However, a significant deviation from the optimal temperature can still exist in radial direction of the tubes due to the radial heat transport resistances as shown in figure 1.

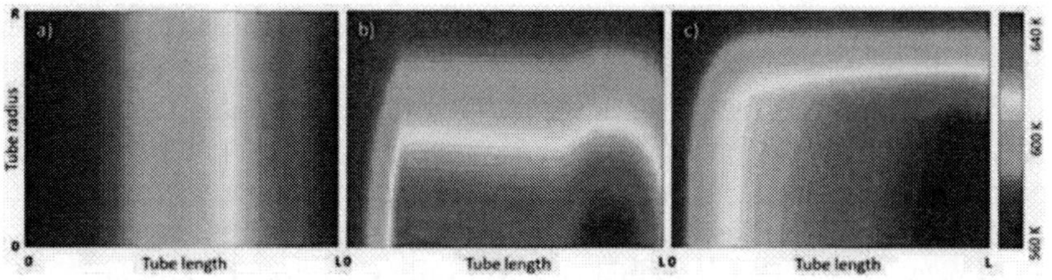

Figure 1: Exemplary demonstration of the (system specific) optimal reaction temperature along the tube (a) compared to the resulting reaction temperature profile due to heat transport resistances in a fixed-bed reactor with an optimal axial catalyst bed dilution profile (b) and a fixed-bed reactor with an optimal axial and radial catalyst bed dilution profile (c) for the air-based production of ethylene oxide.

This deviation can be reduced by further increasing the surface-to-volume ratio of the tubes, thereby reducing the radial transport pathways. To maintain the necessary residence time, this approach requires either longer tubes or an increase of the number of tubes. Both options lead to an increase of investment and operation costs as a result of a higher pressure drop and higher material and/or installation costs.

As an alternative, the concept of the radial optimization of tubular reactors is introduced for the model-based reactor design. This is an extension to the state-of-the art approach of optimizing suitable design variables such as the catalyst bed dilution or the catalyst pellet specifications in axial direction of the tube [3]. The radial coordinate is additionally considered as a degree of freedom for the chosen design variables, leading to a superior approximation of the optimal reaction temperature as shown in figure 1 c). The optimization results give a quantitative indication of the potential reactor performance gain by introducing radial profiles of the chosen design variables. It can then be investigated to which extent these profiles can be technically approximated. Furthermore, the optimization results can indicate the potential of suitable alternate reactor concepts (e.g. catalytic foam reactors or monolithic reactors).

The concept of the radial optimization is presented and illustrated by optimizing a tubular reactor for the air-based production of ethylene oxide in a case study. To reduce the computational effort for the resulting large-scale nonlinear optimization problem, pseudo-homogeneous reactor balance equations are used to describe the reactor. The solutions of the catalyst pellet equations are approximated via analytical approximations [4]. A suitable model basis is identified that is compatible with the dynamic optimization approach, yet considers changes of design variables in radial direction of the tube. The objective function is to maximize the conversion of ethylene while fulfilling other reactor performance constraints in terms of selectivity towards ethylene oxide and space-time yield. Results show that by introducing a 2D dilution profile, the yield of ethylene oxide can be increased by 2 %. Furthermore, the 2D optimization results indicate that a novel alternate reactor concept consisting of a catalytic foam reactor followed by a classical particle packed fixed-bed reactor holds significant potential for process intensification.

References

[1] Peschel, A.; Freund, H.; Sundmacher, K.; Ind. Eng. Chem. Res. 49 (2010) 10535-10548

[2] Eigenberger, G.; Ruppel, W.; Ullmann's Encycl. Ind. Chem. Online ed. Wiley-VCH, Weinheim, 2012

[3] Hwang, S.; Smith, R.; Chem. Eng. Sci. 59 (2004) 4229-4243

[4] Pietschak, A.; Kaiser, M.; Freund, H.; Chem. Eng. Res. Des. (2018) submitted

534776 On-Sun Demonstration of Continuous Redox Driven Solar Thermochemical Hydrogen Production

Abstract:

A sustainable hydrogen (H2) economy based on the low-cost production of H2 from renewable resources has the potential to drastically transform the energy sector. A simple two-step reduction/oxidation (redox) cycle allowing for the production of oxygen (O2) and H2 by solar thermal water splitting (STWS) has been demonstrated on-sun. The H2 and O2 are produced in two separate steps, eliminating the need for high-temperature H2/O2 gas separation. In this redox process, oxygen is generated by the high-temperature reduction of an active metal oxide spinel, forming O-vacancies; followed by a second step in which the vacancies are filled during a high-temperature steam exposure, producing H2 fuel. Although the STWS reaction scheme appears straightforward, its implementation as an economically viable process is extremely demanding. In order to commercially realize this technology, several challenges must be overcome including an energy efficient reactor design, stable reactor and heat exchanger materials, robust redox active particles, and an efficient process for recycling inert gas. Here, we present results showing near-isothermal on-sun H2 production utilizing dual lab-scale fluidized beds operated in which redox cycling produces H_2 continuously. Active hercynite particles were redox cycled at the High Flux Solar Furnace at the National Renewable Energy Laboratory. Production rates of H_2 exceeding DOE project targets and a final deliverable of H2 production were demonstrated. The impact of incident solar radiation and hydrogen production is evaluated during on-sun testing. Results of a techno-economic analysis will be discussed as will the path forward. This work demonstrates the viability of STWS leveraging continuous operation of both reduction and oxidation at near-isothermal conditions within a multi-tube solar thermal cavity reactor. This is the first demonstration of continuous STWS to be reported.

516776 The Mechanism Investigation for Process Intensification of the Esterification Reaction By Microwave Irradiation

Abstract:

Microwave has long been recognized to accelerate the rate of chemical reactions. However, the scientists still clash on the question what is main reason of the microwave intensification effect for this acceleration. Microwave changing the relative volatility of some special mixtures discovered by our previous research may open up new insight. For some reactions that have significantly limited by chemical equilibrium, if the product concentration in reaction system can be reduced through the changing the relative volatility of reaction system, the acceleration of reaction rate can be occurred by the special effect of microwave. In this paper, esterification, a typical equilibrium limitation reaction, has been chosen to determine the intensification effect of microwave. The experimental equipment installs a drainage system to move the water that is one of the reaction products. According to compare the experimental results under microwave and traditional heating, the hypothesis can be verified. Moreover, the main influence factor for reactive conversion, microwave power and reaction temperature were systematically investigated in detail to build the kinetic of this reaction in microwave. Experimental results show that microwave improved drainage can realize the reactive acceleration. The intensification effect of microwave for some reactions also can be explained by the investigation results.

532048 Process Intensification Via Batch-to-Continuous Transition in the Production of Lubricants: Acquisition of Robust Reaction Kinetics

Abstract:

Production of specialty chemicals to-date is conducted almost exclusively in batch or semi-batch reactors. Transition to continuous processing promises strong improvements—so-called "process intensification"—over the century-old batch processing technology with lower unit energy consumption, less waste generation, reduced exposure hazards, smaller process footprint, lower capital investment and more consistent product quality. The transition to continuous processing—already demonstrated widely at the much smaller scale of the pharmaceutical industry—thus offers the prospect of truly revolutionizing the manufacture of specialty chemicals.

As a first step, this transition requires the acquisition of robust information about the process chemistry. While batch reactors are highly tolerant to incomplete knowledge (in particular regarding kinetics), the design and operation of continuous processes requires availability of robust kinetics. However, the detailed chemistry and kinetics underlying current processes are in most cases poorly understood (if not completely unknown). Transition to continuous processing hence requires collection of robust and accurate kinetics as a key prerequisite, including impact of impurities and contaminants which play a significant role in many specialty chemicals processes.

This work presents results from a university-industry collaboration between Lubrizol Corporation and the University of Pittsburgh, under the umbrella of the recently formed RAPID Modular Chemical Process Intensification Institute. It aims to demonstrate the transition from batch to continuous processing focused on lubricant chemistry. Specifically, we present results from a kinetic investigation into the production of succinimide dispersants via reaction of a poly-isobutylene succinimide anhydride (PIBSA) with polyamines. The molecules undergo a condensation reaction to produce a large surfactant molecule with a polar head group and a long, oil soluble polymer tail. A complete family of products (which form the largest volume oil additives) is made by varying molecular weight, structure and stoichiometric ratio of the various reactants. However, the reaction kinetics for this class of reactions are unknown and the acquisition of robust kinetics is hence a crucial first step in the process development for a continuous process. We demonstrate that the combination of online IR spectroscopy as the analytical tool, with careful data evaluation (including model cross-validation techniques and analysis of transient traces) allows for the required acquisition of robust reaction kinetics and the identification of optimal operating windows for a continuous process.

525265 Microscale Engineering of Responsive, Flexible and Reconfigurable Particle Structures

Abstract:

We will discuss the principles of microscale engineering of novel classes of flexible and dynamically reconfigurable assemblies from colloidal particles. Examples of such assemblies include magnetically reconfigurable gel networks, shape-changing microbots, and inks for 3D printing. The key to making such structures is the engineering of directional interactions and of flexible inter-particle bonds. In the first part of the talk, we will discuss how complex magnetic polarization patterns on metallo-dielectric microcubes lead to multidirectional interactions and assembly of reconfigurable microclusters. These sequence-encoded clusters can be reversibly actuated and spatially transported by magnetic fields and can be designed to be self-motile in non-Newtonian media. In the second part of the talk we will describe new types of multiphasic capillary gels from particles bound by liquid bridges. The first gel system is made of filaments from magnetically responsive iron oxide nanoparticles suspended in water-oil systems. The nanocapillary binding results in ultra-high filament flexibility. The second multiphasic system that we introduced is a new class of 3D printing inks consisting of PDMS microbeads, liquid PDMS and water. Owing to the capillary binding, such Homocomposite Thixotropic Pastes (HTPs) can be extruded and shaped on a 3D printer. The curing of the liquid bridges in the HTPs results in remarkably elastic and flexible porous silicone materials, which can be 3D printed under water and find applications such as direct printing of bio-scaffolds.

524022 Tough, High Impact Resistant 3D Printed Objects from Core-Shell Filaments

Abstract:

Fused deposition modeling (FDM), an extrusion-based 3D printing method, uses solid polymeric filaments. Typically, the polymers are formulated to provide desired characteristics and extruded into homogeneous filaments with well-defined diameter. However, an intrinsic weakness of FDM is the poor adhesion between printed layers with the bonding interfaces being poorly developed during printing. These poorly constructed interfaces lead to poor mechanical properties from FDM. Here, we propose a route to attack this intrinsic problem using core-shell structured filaments, where the shell has a solidification temperature much lower (> 100 C) than the core. The disparate solidification temperatures lead to the core solidifying first to act as fiber reinforcement for shape control, while the shell generates a strong shell-shell interface between adjacent filaments. Samples printed from core-shell filaments show enhanced mechanical strength with samples that do not fail and order of magnitude improvements in impact resistance. We discuss how the dimensional accuracy and mechanical properties of the 3D printed part are impacted by polymer selection and print processing parameters.

524105 Solution-Based 3D Printing of Hierarchical Porous Polymers

Abstract:

Current additive manufacturing methods have strong limitations in the classes of polymers that can be printed. This means that many polymers of significant technological interest cannot currently be 3D printed. Here, we demonstrate a generalizable method for 3D printing of viscoelastic polymer solutions and apply it to both porous (a polymer of intrinsic microporosity, PIM-1) and non-porous (cellulose acetate) polymers. Successful ternary ink formulations that result in the creation of three-dimensional structures with hierarchical porosity require balancing of solution thermodynamics (phase separation), mass transfer (solvent evaporation), and rheology. To demonstrate the utility of this approach, a microporous polymer (PIM-1) that cannot be printed with current additive manufacturing technologies is 3D printed into a high-efficiency mass transfer contactor exhibiting hierarchical porosity ranging from sub-nanometer to millimeter pores. Short beds (1.27 cm) of these contactors were challenged with a flow of toluene vapor (1000 ppm) in N_2 gas. These contactors were found to create a pure (< 1 ppm toluene) N_2 stream for 1.7 hours, 6 times longer than traditional structures fabricated from the same PIM-1 material, and more than 4,000 times the residence time of gas in the bed. This solution-based additive manufacturing approach greatly extends the range of materials compatible with 3D printing.

530719 3D Printing with Soft Porous Silicones By the Homocomposite Thixotropic Paste (HTP-3DP) Method

Abstract:

Three dimensional (3D) printing with soft materials is important for a number of applications out of which stand out biomedical materials and scaffolds. One common 3D printing method, direct ink writing, requires thixotropic inks that are flowable at high shear stress but possess high storage moduli and yield stresses. Many precursors for soft materials are liquid at room temperature before crosslinking and are made tixotropic by fillers such as fumed silica, clay nanoplatetes or cellulose nanocrystals. However, the use of fillers is problematic because it may deteriortate the mechanical properties of the resulting composite materials. We will introduce and report the results of a new method of 3D printing of soft materials with homocomposite thixotropic paste (HTP). The new 3D printing inks consist of water and two silicone-based components: crosslinked PDMS microbeads and liquid PDMS phase. Owing to the capillary binding of the microbeads, suspensions containing certain fractions of PDMS liquid precursor behave like pastes, which are flowable at high shear stress but possess the yield stress needed for direct ink writing. The Homocomposite Thixotropic Pastes (HTPs) can be directly extruded and shaped on a 3D printer. The liquid PDMS bridges are thermally crosslinked after printing, resulting in remarkably elastic and flexible structures (Adv. Mater. 2017). Their porosity and mechanical properties, such as tensile modulus, could be controlled by the fraction of liquid precursor in the original multiphasic dispersion. As the HTP-3DP inks are made of porous biocompatible silicone, they can be 3D printed under water and may find applications such as direct printing of bio-scaffolds on live tissue.

534158 Additive Manufacturing of Polypropylene/Hydrogenated Resin Blends: Effect on Crystallinity, Morphology and Mechanical Properties

Abstract:

Additive Manufacturing of Polypropylene/Hydrogenated Resin Blends: Effect on Crystallinity, Morphology and Mechanical Properties

Arit Das[1,+], Alexandra Marnot[1], Eugene Joseph[1] and Michael J. Bortner[1,*]
[1]Department of Chemical Engineering, Virginia Tech, Blacksburg, VA 24061
*Corresponding Author: mbortner@vt.edu (Dr. Michael J. Bortner)
[+]Presenting Author: darit7@vt.edu (Arit Das)
Co-author 1: malex15@vt.edu (Alexandra Marnot)
Co-author 2: sboppfilm@yahoo.com (Dr. Eugene Joseph)

Abstract:
Additive manufacturing (AM) is one of the fastest growing processing techniques in recent times and is fast becoming an alternative to subtractive manufacturing or milling. The most commonly used method is the fused filament fabrication (FFF) due to its operational flexibility, facile maintenance, low energy demands, low material wastage and low cost[1]. However, in spite of notable development in the FFF process, there is still a need to develop and study new materials that can be used for the FFF based AM process. Since becoming commercially available in 1958, isotactic polypropylene (iPP), a semicrystalline polymer, has become a commodity with widespread use throughout nearly all commercial and consumer industries. The crystallization behavior of iPP creates a significant challenge during extrusion based additive manufacturing because it crystallizes relatively rapidly and traps stresses in the layered parts, which in turn causes warpage[2]. Therefore, it is of paramount importance to control the crystallization process in order to successfully 3D print materials using iPP. It has been reported in literature that the addition of low molecular weight hydrocarbon compounds to polypropylene shifts the onset of crystallization to lower temperatures, decreases the rate of crystallization and reduces the overall amount of crystallinity when cooling it from the melt[3].

In this work, blends of iPP with two types of hydrocarbon resins having different degree of hydrogenation have been prepared and the morphology, thermal properties and crystallinity the blends have been studied. In addition, the effect of the amount of resin present in the blend on the above-mentioned properties has been evaluated. Non-isothermal DSC results showed that with increasing resin content the crystallization temperatures of the blends decreased from 121°C (pure iPP) to 117°C. A similar trend was observed for the melting temperature as well which decreased slightly from 165°C to 161°C. The degree of crystallinity decreased from 57% for pure iPP to 42% for the blends due to the addition of partially miscible and non-crystalline resins to pure iPP. The growth of crystal nuclei during the crystallization process was monitored using polarized optical microscope coupled with a hot-stage. The addition of partially hydrogenated resin was found to slow down the crystallization of the blends more effectively than the blends of completely hydrogenated resin. This may be attributed to the presence additional nucleating sites in the fully hydrogenated blends as compared to the partially hydrogenated ones. It is expected that by analyzing the change in morphology and slowing down the crystallization process, we might be able to improve self-diffusion within the blends prior to crystallization, since diffusion is hindered in the crystalline state. Understanding the thermal behavior and morphology of the blends at different temperatures will play a pivotal role in facilitating filament extrusion and subsequent extrusion based additive manufacturing.

References:
1. Ambrosi, A. & Pumera, M. Chem. Soc. Rev. 45, 2740â€"2755 (2016).
2. Coxon, L. D. & White, J. R. Polym. Eng. Sci. 20, 230â€"236 (1980).
3. Cimmino, S., Silvestre, C. & della Vecchia, G. J. Appl. Polym. Sci. 92, 3454â€"3465 (2004).

536378 Filament Extension Atomizer: Novel Aerosol Generation from Polymer Melts and Applications in Additive Manufacturing

Abstract:

PARC, A Xerox Company is developing a novel approach to aerosol generation called Filament Extension Atomization (FEA). FEA is a method that can generate droplets from highly viscoelastic, strain hardening fluids such as molten thermoplastics via a well-studied elastocapillary thinning instability. FEA-generated droplets from these molten polymers can be narrowly-dispersed and could enable multiple disruptive advances in additive manufacturing. PARC aims to develop this key platform technology into multiple use cases in view of additive manufacturing including 1. particle creation for selective laser sintering (SLS); 2. direct ionographic printing of 3D objects and 3. digital deposition of polymer coatings. In particular, since FEA can generate droplets from virtually any thermoplastic regardless of resin type or material grade, PARC FEA has the potential to disrupt polymer-based additive manufacturing by increasing the range of materials that can be used in either existing or upcoming methods. Preliminary data on our research efforts on these use cases will be discussed in this contribution.

537618 3D Printing of Hydrogels with Spontaneous Formation of Solvent-Induced Patterns

Abstract:

Surface patterns driven by instability are commonly observed in nature. Numerous intriguing surface morphologies which provide important functions can be found in living creatures, such as folds in the human brain, wrinkles on human skins, and buckling protrusions on tumor surfaces. Numerous methods have been studied in an effort to produce controllable engineered surface patterns. Yet, majority of the work mainly focus on creating planar surface patterns with limited control on spatial distribution of patterns and not capable of fabricating 3D patterned structures, which significantly limit their applications. In this study, we used extrusion-based direct-ink-writing (DIW) using ink formulations that are able to form a wide range of solvent-induced surface instability patterns in hydrogel form. The inks are formulated from poly(hydroxyethyl methacrylate) (PHEMA) with ethylene glycol dimethacrylate (EGDMA) in the presence of a photoinitiator. We first showed the effect of ink composition and curing process on rheological properties and printability of the inks as well as pattern formation on 3D printed hydrogels. Then, by using multi-material printing we showed spatial control of pattern morphology on 2D substrates. Finally, we demonstrated fabrication of 3D scaffolds/constructs with a variety of ordered micropatterns, including hexagonal and lamellar patterns. Our approach involves three sequential steps in order to obtain optimal ink viscosity and the desired patterns. First a mixture of HEMA monomers and photoinitators were irradiated with UV light (for t_1 second) to obtain a partially polymerized poly(HEMA) prepolymer solution. The solution was then loaded in a syringe and followed by a second round of UV exposure (for t_2 second) in the presence of crosslinkers and additional photoinitiators. The printed object was subjected to UV light for t_3 second after completion of each layer. The finished print was immersed in water for 30 minutes to induced patterns. The three UV exposure times, together with the crosslinker concentrations, govern the surface morphology of the printed scaffolds, offering a novel strategy for extending the use of surface instability patterns.

527001 Study of Local Boiling Heat Transfer for Micro Nano Surface Structures Using a 3D Transient Heat Conduction Model

Abstract:

Corresponding author. E-mail address: hengyi@mail.sysu.edu.cn (Yi Heng)

Inverse heat conduction problems (IHCP) play an important role in both theory and applications. It has a wide range of applications in different fields, such as aerospace engineering, chemical engineering and nuclear engineering [1-2]. In particular, towards the understanding of a complex nonlinear dynamic process - pool boiling, which is commonly used in chemical engineering, efficient 3D IHCP solution techniques in the study of boiling heat transfer mechanism have received more and more attention. Together with the experimental efforts, the reconstruction of surface heat flux distribution and information about two-phase vapor-liquid structures above the surface are indispensable for full understanding of boiling heat transfer in all stages. To this end, many researchers make every effort to modify surfaces to enhance boiling heat transfer [3]. Wang et al. [4] enhanced heat transfer performance by improving the wettability of surfaces. Enhancing boiling heat transfer by using micropores with reentrant cavities and various microchannel geometries was demonstrated by Deng et al. [5] and Walunj et al. [6], respectively.

In order to precisely study the mechanism of enhanced boiling in micro-nano structures, we consider complex geometric domains (inspired by the micro-nano surface structures produced in our recent work [7]) to accurately modeling the 3D transient heat conduction process in pool boiling experiments. We have already analyzed the forward heat conduction problems for flat-plate surface structures focusing on the study of the influence of heater thickness on the inverse solution quality against perturbed temperature measurements [8]. In this work, we want to extend and use the aforementioned analysis method to investigate how honey-comblike porous structures (cf. Figure 1) enhance boiling heat transfer.

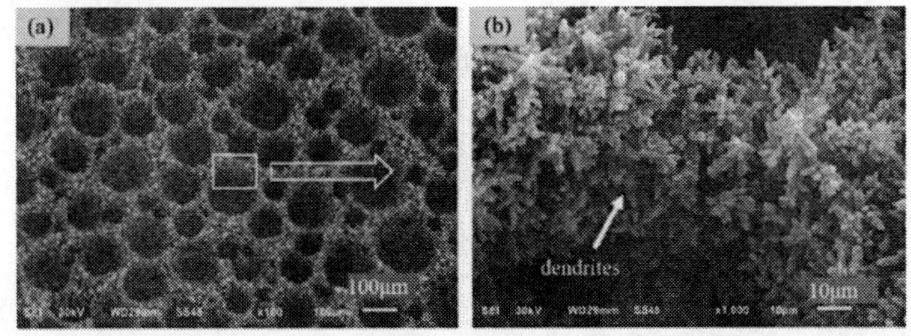

Figure 1 Produced honey-comblike micro porous copper surfaces (adopted from [7])

The 3D transient nonlinear forward heat conduction model defined on the computational domain constrained by the complex micro-nano surface structures is illustrated as follows:

$$\rho(\theta)c_p(\theta)\frac{\partial\theta}{\partial t} = \nabla\cdot\left(\lambda(\theta)\nabla\theta\right), \qquad in\ \Omega\times(0,t_f), \qquad (1)$$

$$\theta(\cdot,0) = \theta_0(\cdot), \qquad on\ \Omega, \qquad (2)$$

$$\lambda\frac{\partial\theta}{\partial n} = q_i, \qquad on\ \Gamma H\times(0,t_f), \qquad (3)$$

$$\lambda \frac{\partial \theta}{\partial n} = q_r, \qquad on\, \Gamma R \times (0, t_f), \qquad (4)$$

$$\lambda \frac{\partial \theta}{\partial n} = q_b, \qquad on\, \Gamma B \times (0, t_f), \qquad (5)$$

where Ω represents the heater geometry. The observation time is [0, t_f] and *n* denotes the outer normal on the boundaries. q_i, q_b and q_r denote the heat fluxes through the heated boundary, the boiling surface , and the rest of the boundaries . , c_p and denote the temperature-dependent density, heat capacity and heat conductivity of the material, respectively. represents the temperature variable and corresponds to the initial temperature.

ΓH

ΓB

ΓR

ρ

λ

θ

θ_0

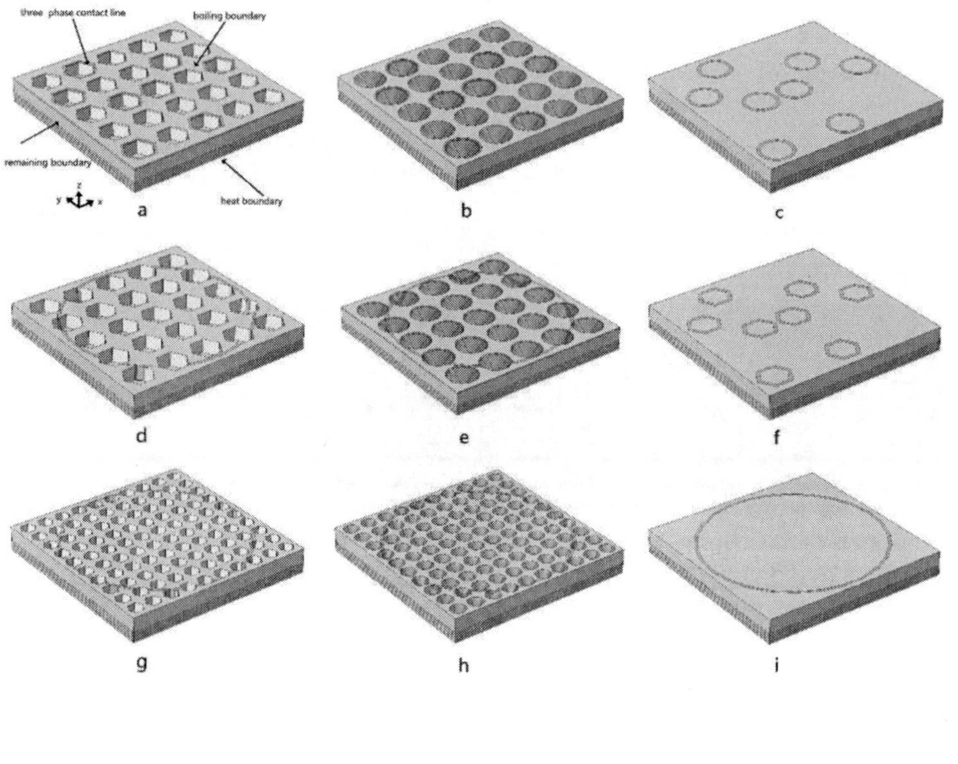

Figure 2 Heater geometries based on the flat-plate surface structure (c, f, i) and two artificial honey-comblike porous structures (hexagonal prisms (a, d, g) and cones (b, e, h)

For the numerical simulations, we consider 9 heaters with the size of 1000um*1000um*130um (see Figure 2). The length of hexagonal micropores are 80um (Figure 2 (a), (d)) and 40um (Figure 2 (g)). The radius of cone micropores are 80um (Figure 2 (b), (e)), 40um (Figure 2 (h)) and the depth of pores are all set to be 75 um. All these geometric settings are based on our recent experiments [7]. Furthermore, according to the micro-layer theory proposed by Stephan [9], most of the heat during boiling is transferred in the micro region of the three-phase contact line (TPCL) by evaporation. Therefore we consider a ring-shaped boiling heat-flux pattern that has been investigated in our previous work [8] for the simulation tests. In Figure 2, the areas of TPCL are marked in blue. The ring width is 16 um, consisting of one part of 8um on the surface and the other part of 8um on the side of hexagonal and cone micropores. The simulated input heat flux q_i at is set to 0.1 MW/m^2 and the

ΓH

simulation time is $[t_0, t_f]$ = [0, 0.05] s. The heat flux out from TPCL complies with

$$\begin{cases} q_o = 5 \times \sin[100 \times \pi \times (t-0.02)] + 5MW/m^2 & on\,\Gamma B \times [0.015, 0.03], \quad (1) \\ q_o = 0 & on\,\Gamma B \times (0, 0.015) \cup (0.03, 0.05), \quad (2) \end{cases}$$

Since the surrounding of heater is covered by aerogel blanket we assume q_r=0. The maximum heat flux out from TPCL is set to 10MW/m^2 following the results from our previous work Heng et al. [10]. In this study, we conduct 9 simulations considering different pore structures, pore density and local heat-flux patterns. The initial temperature for all simulations is set to be 143. While in the cases of Figure 2(a), (b), (c) and (f) we assume TPCLs randomly

°C

occur on the boiling surface that lead to a few small-sized ring-shaped local boiling heat fluxes. A single, large ring-shaped boiling heat-flux pattern is investigated in the cases of Figure 2(d), (e), (g), (h) and (i).

COMSOL Multiphysics is used for the simulations. The finite elements used for all case studies are chosen between 1.6 million and 9 million. The simulation time step is 0.001s. Simulation results of temperature distributions on the heated wall are shown in Figure 3, Table 1 and Table 2, respectively. The order of the results in Figure 3 follows that of the case studies shown in Figure 2. While Figure 3 (a), (b), (c) show the influence of different structures on temperature distributions, Figure 3 (d), (e), (g), (h) show the simulated temperature distributions as the pore density increases.

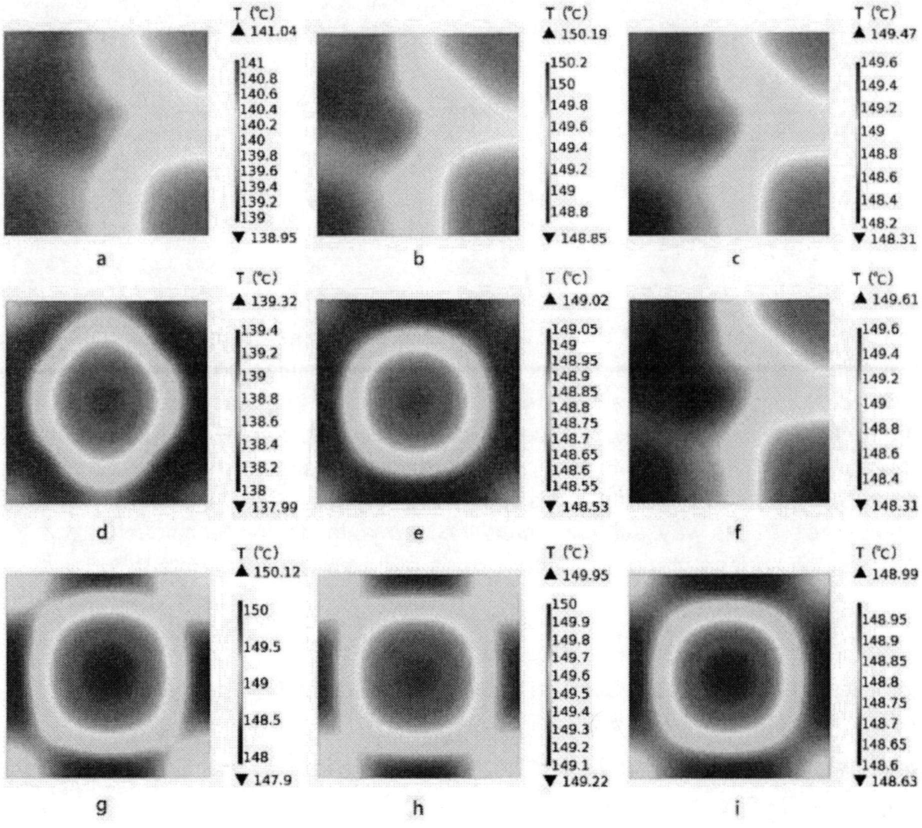

Figure 3 Representative examples of simulated temperature distributions at t = 0.025s

Furthermore, Table 1 and Table 2 summarize the temperature variations on the heated wall in all cases. Due to the increased TPCL areas, at the same level of boiling heat fluxes, the honey-comblike surface structures (Table 1 (a), (b)) lead to higher temperature variations on the heated wall than those on the flat-plate surfaces. As shown in Table 2, we observe that the increasing of pore density for both hexagonal and cone surface structures (Table 2 (g), (h)) lead to even higher temperature variations than those using low-density ones (Table 2 (d), (e)), respectively, which are all higher than those using the flat-plate surface structure (Table 2 (i)). It is concluded that micro-nano surface structures can in general lead to high temperature variations and enhance the local boiling heat transfer. Besides, as discussed in our previous work [8], higher temperature variations in the temperature measurements can also help improve the solution accuracy of IHCP in the presence of same level of measurement error. Hence, the use of micro-nano surface structures shows its advantages.

Table 1 Temperature variations on the heated wall for the simulations of a few randomly distributed small-sized ring-shaped boiling heat fluxes

Temperature(K)	a	b	c	f
Average	412.92	422.56	421.95	421.99
Maximum	414.19	423.34	422.62	422.76
Minimum	412.10	422.00	421.46	421.46
Variations	2.09	1.34	1.16	1.30

Table 2 Temperature variations on the heated wall for the simulations of a single, large ring-shaped boiling heat flux

Temperature(K)	d	e	g	h	i
Average	411.71	421.87	422.16	422.70	421.92
Maximum	412.47	422.17	423.27	423.10	422.14
Minimum	411.14	421.68	421.05	422.37	421.78
Variations	1.33	0.49	2.22	0.73	0.36

In summary, we studied the local boiling heat transfer for micro-nano surface structures using a 3D transient heat conduction model. The effects of selecting different surface structures and pore densities on the temperature distribution results were investigated. Numerical results show the modified micro-nano surface structures can help enhance local boiling heat transfer, which validated our experimental studies [7]. One of our future research will be the study of more complex surface structures containing crystal branches arising from real experiments. Furthermore, we aim to extend our CG-based Iterative regularization method [8] to solve 3D transient IHCP that can deal with complex surface structures. In this way, suitable optimal experimental design techniques are expected to be developed to optimize the surface preparation procedures and boiling experiments to further enhance boiling heat transfer.

REFERENCES

[1] Y. Wang, et al., Effects of anodic oxidation for combustion chamber on heat transfer of the piston in an aero-engine, Experimental Heat Transfer 30(1) (2017) 46-65.

[2] S. Alyokhina, et al., Unsteady heat exchange at the dry spent nuclear fuel storage, Nuclear Engineering and Technology 49(7) (2017) 1457-1462.

[3] D.E. Kim, et al., Review of boiling heat transfer enhancement on micro/nanostructured surfaces, Experimental

Thermal and Fluid Science 66 (2015) 173-196.

[4] Y.Q. Wang, et al., Wettability modification to further enhance the pool boiling performance of the micro nano bi-porous copper surface structure, International Journal of Heat and Mass Transfer 119 (2018) 333-342.

[5] D. Deng, et al., Pool boiling heat transfer of porous structures with reentrant cavities, International Journal of Heat and Mass Transfer 99 (2016) 556-568.

[6] A. Walunj, et al., Comparative study of pool boiling heat transfer from various microchannel geometries, Applied Thermal Engineering 128 (2018) 672-683.

[7] Y.Q. Wang, et al., The Two-Layers Composite Structure of Biomimetic Copper Forest and Honeycomblike Porous Structure to Enhance Pool Boiling Performance, 16th International Heat Transfer conference (submitted) (2018) Beijing.

[8] Y. Heng, Mathematical Formulation and Efficient Solution of 3D Inverse Heat Transfer Problems in Pool Boiling, (2011) RWTH Aachen University.

[9] P. Stephan, et al., A new model for nucleate boiling heat transfer, Heat and Mass Transfer 30(2) (1994) 119-125.

[10] Y. Heng, et al., Reconstruction of local heat fluxes in pool boiling experiments along the entire boiling curve from high resolution transient temperature measurements, International Journal of Heat and Mass Transfer 51(21-22) (2008) 5072-5087.

531617 Novel Adsorption-Based Separation Technology for Gas Treating Applications

Abstract:

Gas processing and treating systems often can be expensive, have a large footprint, and require large utility loads and elaborate maintenance programs. With the prevailing low commodity price environment, there is increased focus on reducing costs associated with these systems. This is particularly important when operating in challenging cost environments such as deep water, arctic, or remote offshore. In response to these challenges, ExxonMobil has been developing a novel adsorption-based gas separation technology with a focus on process intensification. This technology uses rapid pressure- and/or temperature-swing cycles to dramatically reduce equipment size and weight, and overall gas treating footprint by ~50%, when compared to conventional technologies. Furthermore, this technology is well suited for a skid-based deployment approach, which has significant execution efficiencies. The weight and footprint savings have several benefits, in addition to CAPEX savings, particularly in locations where space is at a premium.

The current focus of development is deep dehydration and CO_2 removal for cryogenic applications, such as NGL recovery or LNG. A significant amount of fundamental research and pilot testing has been completed demonstrating technical feasibility. This paper will provide background on the technology platform, the key enablers and overall process integration. Furthermore, this paper will also present data from the pilot studies and the approach adopted for scale-up and demonstration.

533415 Process Intensification, a Promising Approach in Separating a Ternary System Using Distillation to Reduce Energy Consumption: Case Study

Abstract:

Separating a ternary mixture into its pure components is a very often required in most of the processes observed in the various industries. At this moment, Distillation is the commonly used process to separate binary or ternary or multi component mixtures provided enough boiling point difference exist among the components. Though it is a frequently used process, distillation is an energy intensive process, which necessities thorough optimization in terms of energy to meet the profitability and sustainability. As the ternary and more component mixtures require series of distillation columns to get the pure components, a systematic approach to effectively optimize the performance of the distillation units in terms of energy, product purity, and operability, is required. For that reason, employing process intensification, energy integration with thorough thermodynamic analysis is a way forward for a cost-effective and energy efficient distillation column network.

In this study, a systematic energy integration of the distillation separation system is developed to get minimum energy consumption with less capital cost. The optimization is carried out in terms of distillation sequence, process intensification, pressure, temperature & composition of the column, no.of stages, and feed tray location. The extent of the energy reduction depends on the composition of the feed as well. The above-explained approach has been utilized in a real time case, which consumes 20% less energy than that of presently employed conventional distillation method for the chosen ternary system. This new method also requires less capital cost for a green field project. The paper also advises general heuristics to employ a particular distillation sequence for separating a ternary mixture system. The methodology employed in the study, can also be utilized to optimize four-component mixture systems.

535443 Improved Fixed-Bed Transport Characteristics: A Shortcut Method to Optimize Catalyst Pellet Specifications

Abstract:

Highly efficient reactors are fundamental for the resource- and energy efficient production of chemicals. The basis to develop such optimal chemical reactors is the simultaneous consideration of all possible design variables, leading to a perfectly tuned system. In catalytic fixed-bed reactors, the catalyst pellet specifications have a significant impact on the system behavior. Besides influencing the catalytic activity and selectivity due to intraparticle transport phenomena, the pellet also influences the fixed-bed transport characteristics such as heat transfer and pressure drop. Hence, the simultaneous design of reactor and catalyst pellet can lead to an improvement of the overall system performance.

For model-based reactor design via dynamic optimization [1] a rigorous treatment of the catalyst pellet during reactor design demands the discretization of the pellet balance equations in addition to the reactor balance equations. The result is a complex heterogeneous model leading to challenges in regard to the computational effort and the numerical stability during optimization. Alternate approaches exist that enable a consideration of the intrapellet mass transport effects via analytical equations. However, these approaches either demand extensive pre-calculations (e.g. Adomian decomposition method [2]) or are strongly limited in regard to their applicability to certain reaction systems or reaction kinetic models (e.g. Thiele-modulus concept [3]). To avoid the drawbacks of these approaches, in this contribution a novel shortcut method is developed and applied. This approach is applicable to arbitrary reaction networks and reaction kinetic models. It enables the computationally cheap optimization of catalyst pellet specifications to improve the bed transport characteristics during the simultaneous design of reactor and catalyst pellet.

The principle of the method is to linearize the rate of formation of each component influencing the reaction rates at bulk conditions. These linear expressions are inserted into the corresponding catalyst pellet component balances. The resulting set of identically structured second order inhomogeneous differential equations is parametrized, decoupled and then solved analytically. The solutions, differing only in the values of their parameters, yield the component concentration profiles in the pellet in dependency of the pellet specifications. Effective reaction rates, calculated via integration, can be inserted into the reactor balance equations. Hence, pellet specifications can be optimized without the need of a heterogeneous model. To ensure a high accuracy of this approach, high catalyst efficiencies should be targeted.

To demonstrate the potential of a simultaneous reactor and catalyst pellet design with the developed shortcut method, a fixed-bed reactor with constant coolant temperature for the production of ethylene oxide is optimized. The objective is to minimize the pressure drop while fulfilling other reactor performance constraints. Pellet specifications in terms of pellet size and pellet heat conductivity are optimized. A pressure drop reduction of 50 % compared to an optimized reactor with prefixed pellet specifications is achieved. Figure 1 shows the optimized catalyst heat conductivity as well as the resulting reaction temperature with reaction progress.

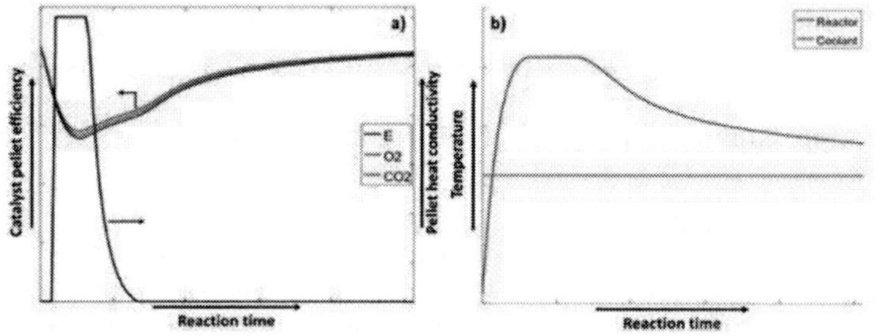

Figure 1: Left: Optimal catalyst pellet heat conductivity profile and resulting catalyst pellet component efficiencies (E=Ethylene, O2=Oxygen, CO2= Carbon dioxide). Right: Resulting reaction temperature profile and optimized coolant temperature.

The presented method enables the design of tailored reactor-catalyst pellet systems to improve the bed transport characteristics. Results with respect to the pellet specifications can point to certain catalyst pellet types available and/or give impulses to develop pellets fulfilling the optimal specifications, leading to highly efficient reactor systems.

The approach developed in this work is generally applicable, i.e., no limitations exist regarding the applicability to certain reaction networks or specific types of reaction rate models as compared to existing methods. At the same time, since a high model complexity is avoided, the approach can be used for efficient simulation and optimization of complex reaction systems.

References

[1] Peschel, A.; Freund, H.; Sundmacher, K.; Ind. Eng. Chem. Res. 49 (2010) 10535-10548

[2] Adomian, G.; J. Math. Anal. Appl. 135 (1988) 501-544

[3] Thiele, E.W.; Ind. Eng. Chem. 31 (1929), 916-920

[4] Aris, R.; Chem. Eng. Sci. 13 (1960) 18-29

537471 Multi-Material, Microchannel Heat Exchanger Design for Enhanced Heat Exchange Processes By Anisotropic Conduction

Abstract:

Recent advances in rapid prototyping techniques, namely laser powder bed fusion, present an opportunity for intensification of heat exchange processes through anisotropic conductive heat flow. Multi-material microchannel heat exchangers â€“ which exploit highly conductive, micro-pillar architecture â€“ are explored here for enhanced rates of heat transfer in the direction of interest only. By reducing heat flow in undesired directions and decreasing the characteristic time for heat transfer between the fluids, it is shown that a reduction of residence times and heat-exchanger volume can be achieved. For a variety of cases, volumetric heat transfer coefficients for systems with and without conductive micro-pillars are evaluated numerically and compared. A 3D printed prototype is also built and tested to demonstrate the functionality of multi-material heat exchangers and to verify numerical results.

538723 Small-Scale Production of Platform Chemicals from Coal with Low-Temperature Microwave Plasma

Abstract:

The United States is endowed with extraordinary riches: it holds more than a quarter of the world's coal reserves. Regrettably, this wealth is severely undervalued. Although coal has catalyzed industrial development in the 19th and early 20th century not only as a ready source of energy, but also as the principal source of synthetic chemicals and materials â€" from aspirin to benzene to naphthalene to tarmac, â€" over the past decades coal became essentially synonymous with production of low-cost electricity.

Shale revolution has resulted in an abundance of low-cost natural gas and prompted the fall of coalâ€™s share in electricity production from 50% to less than 30% over the course of a single decade. With significantly depressed domestic thermal coal prices, there is an opportunity to reorient the chemical industry back to reliance on this accessible and abundant feedstock. It has already happened in China, for instance, where the present capacity of coal gasification/Fischer-Tropsch conversion to methanol targeting production of olefins is 13.6Mtpa, with $55 billion worth of new capacity planned through 2026. FT plants, however, have conventionally been associated with outsized CAPEX investment (in billions USD) and a large environmental footprint, precluding their adoption in the United States.

A viable alternative method for production of coal-derived platform chemicals is direct, thermal conversion (pyrolysis) of coal. Known and industrially employed since 1800s, destructive distillation of coal yields hydrocarbon gases, liquids (monoaromatics and coal tar), and coke. In fact, until 1950s, the coke ovens were responsible for virtually 100% of the domestic benzene supply, but were quickly displaced by petroleum refiners. While the first benzene produced from petroleum appeared only in 1950, by 1961 petroleum benzene accounted for more than 75% of the total production.

Today, benzene and other monoaromatics (BTEX) are the crude oil-derived platform chemicals at the base of many important industrial supply chains: plastics, rubbers, fabrics, solvents, detergents, dyes, and pharmaceuticals. The global demand for benzene alone is in the order of 40 million tons per year. Domestic benzene deficit is expected to grow from 1.8 million tons in 2014 to 4.4 million tons by 2023, due to impact of the shale revolution on domestic BTEX sources. Refineries are forced to run reformers at reduced rates, while BTEX streams from naphtha cracking are reduced as low-cost high-ethane shale gas becomes preferred feedstock for ethylene production.

H Quest has developed and demonstrated a novel coal conversion process that avoids heat transfer and scale requirements limitations of conventional coal conversion technologies, and is well-suited for small-scale, distributed production of chemicals and material precursors from coal. In place of conventional thermal inputs, this rapid, continuous process uses focused microwave energy to create non-equilibrium plasma within inert or hydrogen-rich (CH_4, H_2 or syngas) process gas, localized around entrained coal particles. Optical (spectrometry) diagnostics are used for reaction characterization, species identification, and temperature determination. Observed intensity ratios or spectra band shapes yield temperature by Boltzman analysis. Pyrometry measurements of the solid particles falling through the MW reaction zone yield a temperature of ~ 1200 K. The entraining gas remains relatively cool: thermocouple measurements at the outlet of the microwave reactor typically do not exceed 300 C. Thus, rapid (< 1 sec) pyrolysis of coal releases liquid products and intermediates into the relatively cool process gas, where immediate quenching prevents secondary cracking reactions. The result are the high yields of the relatively high-value liquids (50-60wt%, d.a.f) and minimal yields of the low-value gases (<5wt%).

Process gas composed of methane or syngas serves as a hydrogen source within the single-stage reactor, eliminating external hydrogen production units and the associated CO_2 production, water consumption, and capital costs. Its ionization within the localized reaction zones produces active CH_x* and H* radical species, which promote further coal decomposition, prevent retrogressive radical recombination reactions of oil intermediates, and improve quality of the liquid products through partial hydrogenation and alkylation reactions. Direct activation of a hydrogen-rich gas within localized reaction zones eliminates requirements for high pressures, which conventional coal-to-liquids processes relies on to overcome mass transfer limitations.

Thus, the process enables deployment of a continuous, small-footprint, high-throughput conversion system with *a thermal efficiency approaching 80% and life-cycle CO_2 emissions and water consumption at least 50% lower than conventional coal conversion technologies.*

Varying energy input, feed rate, and process gas composition allows adjusting the liquid product composition and properties. Bituminous coals processed in inert atmosphere (process gas includes inert, recycleable argon) yield highly aromatic liquids. Up to 25% of the liquids are benzene and other monoaromatics. Co-processing with activated hydrogen-donating gases results in methylation, hydrogenation, and ring-opening of the aromatic compounds, as shown by GC/MS traces of the resulting products and corroborated by optical spectrometry. Experimental data from testing high-sulfur (>3%) bituminous coals from Illinois and Ohio shows reduction of sulfur content relative to the parent coal in both char and liquid products of 50% and 80% respectively, effectively beneficiating both products. Production of excess hydrogen from methane decomposition may be used in downstream processes or sold as a valuable by-product.

In the past, coal conversion processes were hindered by inefficiencies, vast capital costs, high environmental impact, and, as a result, by non-competitive costs of production. New technologies relying on unconventional pathways and methods can enable cost-effective, small-scale, clean, efficient value-added production of high-value chemicals and materials directly at the coal mine sites. Most importantly, deployment of these technologies would have a direct and rapid impact on job creation and economic development in the rural communities most affected by the recent downturn in the coal industry.

Acknowledgements

H Quest Vanguard, Inc. is a privately held technology company, based in Pittsburgh, Pennsylvania, focused on the development and commercialization of novel hydrocarbon conversion technologies. Work supported by the Department of Energy, Office of Science under award DE-SC0015895 and by the Illinois Clean Coal Institute under subgrant 15-01 have contributed to materials presented herein.

538809 Effectiveness Factor Phenomena for the Transition between PBR (Packed Bed Reactor) and MR (Membrane Reactor) Via Coupled Heat and Mass Transfer

Abstract:

The study of multiphase transport/reaction phenomena is, of course, of interest to a wide range of separation and reactor applications involving of simultaneous heat, momentum, and multi-component mass transport. The most marked characteristic of such multiphase systems is their nonlinear nature, which requires the quantification of a multiple-coupled set of nonlinear differential equations for the problem solutions that entails a complex set of technical challenges. For reactors, including the MR studied here, accounting for all such phenomena dictates the use of the so-called heterogeneous models rather than the simplified pseudo-homogeneous models that have, to date, dominated the MR modeling literature. This work focuses on the multi-scale modeling of a membrane reactor (MR within which the water gas shift (WGS) reaction is carried out. The work assesses the MRâ€™s effectiveness in transforming, coal-derived syngas into two streams, one rich in hydrogen and the other rich in carbon dioxide. It is shown that significant variation of the catalyst pellet effectiveness factor occurs along the reactorsâ€™ axial direction, and that catalyst pellets of the same diameter exhibit different effectiveness factors within the PBR and MR environments. The effect of catalyst solid thermal conductivity on temperature gradients within the catalytic pellet, and the effect of sweeping gas pressure/temperature on the MRâ€™s behavior, are both quantified. Adiabatic and wall-isothermal operation are shown to exhibit significant differences from each other for both PBR and MR reactors.

532317 Designing Difficult-to-Cyberattack Process Control Systems

Abstract:

Cybersecurity of chemical process control systems is critical because attacks on these processes can harm physical equipment as well as people [1]. Cyberattacks can take many forms, including sending false sensor measurements to operators in an attempt to make them think that the process is operating normally when it is not [2] or manipulating the control actions applied to the process [3]. In [4], it is noted that physical attacks on plants may be prevented by making the plants less attractive to attack by reducing the consequences of an attack by making a plant more inherently safe (e.g., using less hazardous materials); a similar concept should apply in the context of cybersecurity.

Motivated by the above considerations, we develop an implementation strategy for model predictive control (MPC) that seeks to make the control system more difficult to cyberattack, when the specific type of attack considered is one in which the attacker provides false state measurements to the MPC. We develop an implementation strategy in which a variety of control designs with stability-based constraints (e.g., constraints based on Lyapunov and Control Lyapunov-Barrier functions) are developed and one is randomly selected at every sampling time. Each controller on its own can guarantee closed-loop stability and recursive feasibility for initial states within a well-characterized region of state-space, and the implementation strategy ensures that the only controllers available at a given sampling time to be selected between are those which would be feasible, such that the implemented control actions guarantee closed-loop stability even when randomly selected. The goal of the random selection is to make it difficult for a cyberattacker to determine which incorrect value of the measured state to provide to the controller at a given sampling time to achieve a certain goal since the controller which will be used with that measurement, and therefore the input to be applied when the state measurement is provided, is difficult to discern *a priori*. A chemical process example demonstrates the use of this randomized MPC implementation strategy.

[1] A. A. Cárdenas, S. Amin, Z.-S. Lin, Y.-L. Huang, C.-Y. Huang and S. Sastry, "Attacks against process control systems: Risk assessment, detection, and response," In *Proceedings of the ACM Asia Conference on Computer & Communications Security*, Hong Kong, China, 2011.

[2] O. Linda, M. Manic and M. McQueen, "Improving control system cyber-state awareness using known secure sensor measurements," In *Proceedings of the International Conference on Critical Information Infrastructures Security*, Lillehammer, Norway, 2012.

[3] A. Roisch, H. Voos, Y. Li and M. Darouach, "A model predictive approach for cyber-attack detection and mitigation in control systems," In *Proceedings of the IEEE Conference on Decision and Control*, pages 6621-6626, Florence, Italy, 2013.

[4] S. Bajpai and J. P. Gupta, "Site security for chemical process industries," *Journal of Loss Prevention in the Process Industries*, 18, 301-309, 2005.

526412 Cyber Security of Model Predictive Control Systems for Chemical Processes

Abstract:

Recently, the security of process control systems has become crucially important since control systems are vulnerable to cyber-attacks, which is a series of computer actions to compromise the security of control systems (e.g., integrity, stability and safety) [1]. Among cyber-attacks, targeted attacks are severe threats for control systems because of their specific designs with the aim of damaging the control actions applied to a chemical process (for example, Stuxnet worms aims to modify the data sent to a Programmable Logic Controller) [2]. Additionally, targeted attacks are usually stealthy and difficult to detect using classical detection methods since they are designed based on some known information of control systems (perhaps they make explicit use of the process state measurement). Therefore, designing an advanced detection system [3] and a suitable model predictive control system for nonlinear processes in the presence of targeted cyber attacks is an important open issue.

Motivated by this, in this work, we develop a model predictive control system for nonlinear systems [4] subject to targeted cyber attacks. Specifically, a cyber attack that aims to damage close-loop stability via a sensor tamper is first considered and applied to the closed-loop process, and it is demonstrated that if not suitable action is taken by the model predictive controller the closed-loop system loses its stability. Subsequently, an anomaly-based detection method is developed using a priori knowledge of control system, and a suitable model predictive control method is developed to reduce the impact of cyber attacks and re-stabilize the closed-loop system in finite time. A chemical process example is used to demonstrate the applicability of the proposed approach.

[1] Ye N, Zhang Y, Borror C M. Robustness of the Markov-chain model for cyber-attack detection. IEEE Transactions on Reliability, 2004, 53: 116-123.

[2] Cárdenas A A, Amin S, Lin Z S, et al. Attacks against process control systems: risk assessment, detection, and response. Proceedings of the 6th ACM symposium on information, computer and communications security. ACM, 2011: 355-366.

[3] Ozay M, Esnaola I, Vural F T Y, et al. Machine learning methods for attack detection in the smart grid. IEEE Transactions on Neural Networks and Learning Systems, 2016, 27: 1773-1786.

[4] Mhaskar P, El-Farra N H, Christofides P D. Stabilization of nonlinear systems with state and control constraints using Lyapunov-based predictive control. Systems & Control Letters, 2006, 55: 650-659.

542728 Keynote Talk: Technology Development for Cybersecure Fossil Power Generation

Abstract:

Distributed sensor networks, secure sensor data transmission, and automated controls are key elements to better power plant flexibility, performance, and reliability. The prevalence of wireless sensors is likely to increase dramatically, supported by Internet of Things (IOT) infrastructure, and driven by a need to use data analytics approaches for condition-based monitoring and fault detection to reduce forced outages, extend asset life, and maximize plant productivity. Additionally, many utilities are leveraging a centralized monitoring and diagnostic approach, either in-house or through a vendor service, aggregating data streams associated with many power plants requiring remote access to plant data. These technologies need to be developed and integrated in a manner that doesnâ€™t create additional system vulnerabilities â€“ particularly related to cybersecurity. The National Energy Technology Laboratory is sponsoring research on behalf of the US Department of Energyâ€™s Office of Fossil Energy to identify and mitigate cybersecurity vulnerabilities related to fossil-based power generation. A review of preliminary cybersecurity vulnerabilities will be presented along with an update on pertinent cybersecurity R&D underway.

543156 Keynote Talk: Cybersecurity Realities in Industrial Control System Environments

Abstract:

Cybersecurity, particularly in the Industrial Control System (ICS) discipline, is hardly a rare topic these days. There are endless discussions of IT/OT convergence, cybersecurity frameworks, IIoT, the growth of Advanced Persistent Threats, and myriad similar topics. However, there are essential differences between the traditional IT approach to cybersecurity and the paradigm that exists in an ICS environment that create unique challenges and provide likewise unique opportunities. This keynote will highlight some of the less obvious aspects of defending these environments and engage the audience to think, perhaps, a little differently. The talk will cover the organizational capability, governance, and philosophical distinctions encountered in real-world cybersecurity deployments and some of the lessons learned by major Oil, Gas, and Chemical producers. Additionally, this session will consider some of the disruptive technologies that have emerged as a result of the rapid expansion of the ICS cybersecurity market. This particular landscape is in rapid transition and it is important to understand some of the esoteric approaches necessary for successful cybersecurity development early in the journey.

526738 Intensifying Natural Gas Upgrading: Integrated Reactor Concepts for Syngas Production

Abstract:

The recent surge in economically accessible natural gas reserves has revived interest in the development of novel, less capital intensive, more resource efficient, and cleaner processes for natural gas conversion via upgrading to valuable chemical feedstocks. This development, often summarized under the label "Process Intensification" (PI), is based on a conceptual approach to process engineering that aims to radically reduce the physical, energetic, and/or environmental footprint of existing processes without altering the basic underlying process chemistry.

This presentation will discuss examples of process intensification principles applied to natural gas conversion, drawing on work from our own laboratory and others. A particular focus will be on the use of multifunctional reactor concepts, ranging from heat-integrated reactors to chemical looping. I will argue that multifunctionality is at the heart of PI, marking a significant departure from the "unit operations" approach that has defined chemical engineering since its inception in the early 20th century. Systematic application of PI can drastically increase efficiency, process safety, and reduce the overall footprint of next-generation natural gas conversion processes.

527943 Process Concept for Isolation of Low Molecular Weight Carboxylic Acids from Dilute Aqueous Feed

Abstract:

Market launch of biobased bulk products, such as low molecular weight carboxylic acids, alcohols and ketones, very often suffers from poor economics. Low concentration of target constituents is a major obstacle. Process intensification is a promising approach for addressing the economic feasibility of biobased bulk products.

Liquid-liquid extraction with chemical reaction in emulsified state offers excellent technological features for application in this challenge. This concept, successfully applied in multiphase homogeneous catalysis, was investigated for acetic acid isolation by esterification from dilute aqueous effluents with feed concentration of less than 60 g L^{-1}. Conversion rates of up to 55%, compared to conversion of just 4% esterification without catalysis were achieved. For further optimization of this process concept, investigating parameter interaction was necessary to maximize conversion. Design of experiments (DoE) was applied to evaluate relevant process parameters temperature, stoichiometric ratio and catalyst load in batch experiments. Evaluation of design of experiments revealed a possible mass transfer limitation for low phase ratio. Based on the DoE results a model was fitted with high reproducibility and agreement with experimental data. Accuracy of conversion prediction with the model was experimentally verified and confirmed. Downstream processing via transesterification by reactive distillation was applied to complete the technology concept.

*n*â€‘octanol, diluted in the solvent *nâ€‘*undecane, was used as shuttle reactant. *n*-octanol was chosen because of its negligible water solubility. The anionic surfactant 4â€‘dodecylbenzenesulfonic acid catalyzed the esterification reaction. Emulsification during conversion provided large mass transfer area. Regeneration of the shuttle reactant *n*-octanol was performed by transesterification with short chain aliphatic alcohols (e.g. methanol) after emulsion splitting by centrifugation. The low molecular weight products methyl acetate/methanol can simply be separated from the reaction broth by single step distillation due to the huge boiling point difference. It has to be mentioned that the bottom stream of the distillation step may be recycled to the esterification without temperature adjustment. Reaction kinetics of esterification and transesterification for varying stoichiometric ratio and operation temperature level were subject of detailed analysis. For validation, the process concept was further tested for the low molecular weight carboxylic acids formic acid and propionic acid.

537987 Process Integration Using Block Superstructure

Abstract:

Process integration enables the design of integrated production systems ranging from individual processes to total sites to reduce fresh resource consumption or harmful emissions [1-4]. A classic process integration problem consists of sinks with demand for certain properties, and sources to satisfy the demands of these sinks. These properties exchanged within the system can be materials/contaminants in water networks [5], mass exchange networks [6], etc. The properties can also be exchanged as energy, e.g., heat in heat exchanger networks [7], power in work exchange networks [8]. Capturing these recurring features helps to construct a unified superstructure representation for different integration problems. Besides, there is a symbiotic relation among process integration, synthesis and intensification [9-11], which necessitates the advancement of process integration techniques. Recently, the building block superstructure has been proposed for systematic process intensification by dissecting unit operations into fundamental phenomena building blocks [9]. Later on, this approach is applied for general process synthesis problems without postulating case-specific process superstructures [10].

In this work, we extend the block representation to incorporate various process integration problems by considering their common features. Process units, including regenerators/interceptors, are embedded in blocks. The feed flow and product flow in each block represent sources and sinks, respectively. Adjacent blocks interact with each other through direct connecting streams equipped with heaters/coolers for adjusting the temperature. Besides, we also introduce 'jump flow' among all blocks to avoid more intermediate blocks connecting nonadjacent blocks while maintaining or increasing the number of process alternatives. The block superstructure size is dependent on the number of layers with mixing operations, process units, product streams and heat integration stages. We formulate the general process integration problem as a mixed-integer nonlinear optimization (MINLP) problem to minimize the total annual cost. With the proposed MINLP model, we investigate various process integration problems involving mass integration, heat integration, simultaneous mass and heat integration, and property integration. These case studies showed that the same reported optimal integration networks and/or better network designs can be generated using the proposed block representation approach.

References:

[1] Li J., Demirel S.E., Hasan M.M.F. Process Integration using Block Superstructure. *Industrial & Engineering Chemistry Research*, 2018, 57: 4377–4398.

[2] El-Halwagi, M. M. Pollution prevention through process integration. *Clean Products and Processes*, 1998, 1: 5–19.

[3] Friedler, F. Process integration, modelling and optimization for energy saving and pollution reduction. *Applied Thermal Engineering*, 2010, 30: 2270–2280.

[4] Li J., Demirel S.E., Hasan M.M.F. Fuel Gas Network Synthesis Using Block Superstructure. *Processes*. 2018, 6(3): 23.

networks. *AIChE journal*, 2011, 57: 434–457.

[6] El-Halwagi, M. M.; Manousiouthakis, V. Synthesis of mass exchange networks. *AIChE journal*, 1989, 35: 1233–1244.

[7] Yee, T. F., Grossmann, I. E. Simultaneous optimization models for heat integration II. Heat exchanger network synthesis. *Computers & Chemical Engineering*, 1990, 14: 1165–1184.

[8] Razib, M., Hasan, M. M. F., Karimi, I. A. Preliminary synthesis of work exchange networks. *Computers & Chemical Engineering*, 2012, 37: 262–277.

[9] Demirel, S. E., Li, J., Hasan, M. M. F. Systematic process intensification using building blocks. *Computers & Chemical Engineering*, 2017, 105: 2–38.

[10] Li, J., Demirel, S.E., Hasan, M.M.F. Process Synthesis using Block Superstructure with Automated Flowsheet Generation and Optimization. *AIChE Journal*, 2018, under review.

[11] Baldea, M. From process integration to process intensification. *Computers & Chemical Engineering*, 2015, 81:104–114.

536669 *Solids Metrics for Successful 3D Printing of Energetic Feedstocks*

Abstract:

E&G Associates, Inc.

100 Cherokee Blvd, STE 332

Chattanooga, TN 37405

[1] Presenting Author, brandon.ennis@powdernotes.com, (615) 823-0054

[2] Corresponding Author, benjamin.ennis@powdernotes.com, (423) 544-7298

Abstract:

Traditional manufacturing methods for producing warheads, propellants, and pyrotechnic systems are timely and costly. Additive manufacturing (AM) approaches offer one path of alternative manufacturing technologies. Among various technologies, powder bed fusion (PBF) is one promising approach. A PBF printer is essentially a miniature manufacturing facility in a box, with significant powder handling challenges that must be overcome for robust production performance and part quality [1, 2]. Feedstock handling in a powder bed fusion printer is no different than and faces the same challenges as full-scale solids handling processes.

For a feedstock to function properly in a powder bed fusion 3D printer, it must be designed to meet several material characteristics. Some of the most important material properties include: powder flowability & cohesion, powder and roll friction, permeability, particle size & shape distribution (PSSD), triboelectric charging propensity, wetting characteristics, compaction characteristics and thermal softening point.

This paper presents select studies of developing feedstocks for PBF printing of energetic and other materials. Key attributes benchmarked against included powder flowability, aeration attributes, wetting behavior, spreadability, and fusion, as well as final part performance (bulk density, detonation, sensitivity, and part tolerances.)

[1] Ennis, B.J. *et al.* (2008) Section 21: Solids-Solids Operations & Processing. *Perryâ€™s Chemical Engineersâ€™ Handbook*. Perry, R. H., & Green, D. W. (eds.), New York: McGraw-Hill.

[2] Ennis, B.J. et al. (2017), HPC4Mfg Modeling of Powder Dynamics in Metal Additive Manufacturing, Austin, TX. Link for presentation:

https://hpc4mfg.llnl.gov/pdfs/events/powder2017/Powder_Dynamics_EG_17oct2017_FINAL.pdf

[3] Ennis, B.J. et al. (2017). Navy SBIR Phase I Award. Development of Explosive Feedstock for Commercial-off-the-Shelf (COTS) 3D Printers. Link for award:

https://www.navysbirprogram.com/NavySearch/Award/award.aspx?pk=869CC83B-5945-4E75-A598-AA2527D51045

[4] "*Chattanooga firm to help U.S. Navy figure out how to 3-D print explosives*" (2017), T. Omarzu,

Chattanooga Times Free Press. Story link:

http://www.timesfreepress.com/news/business/aroundregion/story/2017/dec/10/chattanoogfirm-help-us-navy-figure-out-how-3-/458777/

536262 3D Printing of Metal-Polymer Composite Structures Via Fused Deposition Modeling

Abstract:

This work demonstrates the ability to 3D print a fluoropolymer based energetic material that can be used as a structural reactive. The developed filament consists of a fluoropolymer binder (polyvinylidene fluoride, PVDF) with embedded aluminum particles. The filament can be printed with commercial 3D printers to create geometrically complex, three dimensional structures. The printing performance of the energetic filament was found to be comparable with standard 3D printing materials, such as ABS. The reactivity of the 3D printed energetic material was shown to be unaffected by the printing process, as determined by comparing the burning rates, as well as differential scanning calorimetry (DSC) and thermal gravimetric analysis (TGA) of the printed material with the unprinted filament. Quasi-static mechanical testing demonstrates this materialâ€™s strength and ability to function as a structural reactive. The original Al/PVDF material was shown to have a higher mechanical strength than PVDF alone. The metal content was also varied in order to obtain relationships between the energetic performance and mechanical strength based on the Al content in the material.

531868 Polymer Resin Systems for Precision Direct-Ink-Write Printing of Thermite-Loaded Inks

Abstract:

Additively manufactured sample preparation allows control of the material scale and hierarchical structure, and can potentially enable unprecedented tailoring of reaction characteristics in energetic materials. Direct-ink-write 3-dimensional printing offers the simplest test-bed to explore numerous material classes. However, the inks for such printing require specific, predictable properties such as shear thinning to accommodate nozzle extrusion, consistent yield stress point and viscosity for predictable flow, and a rapid transition rate between viscoelastic solid and liquid phases. Most classes of energetic materials such as explosives, propellants, and pyrotechnics are solids or composites comprised of highly-solids loaded polymer systems. Optimizing a desired energetic formulation for print quality while also targeting maximum energy density is problematic due to solid-solid interactions dominating the rheology of the ink. Solid-solid interactions make the flow and rheological properties less ideal for extrusion-based printing methods, and require sufficient rheological tunability to be built into the low weight percent reserved for the resin system. The halting of flow after deposition can be greatly improved by rapid, on demand chemical curing mechanisms initiated by light or heat, but such systems are poorly suited as phlegmatizing agents for sensitive energetic materials. An ideal resin system for energetic materials would have sufficient rheology and yield stress tunability to be precision printed with many types of solids, and cure on demand to expand the print design space.

We report a general, dual-polymer-based composite resin system with dual-curing mechanisms that affords extensive adaptability across multiple applications. This resin system allows for precise, omnidirectional printing of highly solids-loaded inks. Using this resin system, we were able to print single walled filaments, overhangs, spans, and a freeform helix. Our work on this family of inert ink-composites with 68-80% solids loading exhibited independently tunable (i) cure rate, (ii) rheology, (iii) glass transition temperature, (iv) print-ability, and (v) tensile/compression properties. Our inks are demonstrated to print as well as current state-of-the-art inks with little to no solids loading. A reactive ink comprised of Al and CuO solids forming a thermite was prepared and characterized. Burn characteristics of the Al/CuO thermite ink as a function of geometry will be discussed.

This work was performed under the auspices of the U.S. Department of Energy by Lawrence Livermore National Laboratory under contract DE-AC52-07NA27344.

532194 3D Printing of Thermite Mixtures Using Static Mixing

Abstract:

Water-based solids-loaded inks are a general solution for 3D printing of functional materials, including reactive materials. While pre-mixed reactive materials inks are possible, it is desirable from a safety standpoint to formulate single-component inks which each contain only one reactant. These inks are essentially inert while separate and only achieve reactive properties when mixed, minimizing the quantity of reactive material present at any one time. We have been exploring various approaches to on-the-fly mixing of such single-component inks for 3D-printed reactive materials. In this talk, I will focus on Al/CuO thermites printed from single-component inks with on-the-fly "static" mixing. Commonly employed for two-part epoxies or dental materials, static mixers work by repeatedly "folding" two fluids together until they are intimately mixed. This is a useful alternative to dynamic mixing using a rotating paddle because it minimizes the energy input into the materials and thus the risk of premature reaction. The mixing quality is also less sensitive to variations in the printing speed. Our initial work has focused on demonstrating feasibility and identifying how much mixing, as characterized by the number of mixing elements, is required to achieve maximum thermite reactivity. After establishing the critical mixing conditions, we have explored some basic 3D-printed structures and begun to assess the impact of architecture on overall reactivity.

This work was performed under the auspices of the U.S. Department of Energy by Lawrence Livermore National Laboratory under Contract DE-AC52-07NA27344. LLNL-ABS-733101

534724 Additive Manufacturing of Pyrotechnic Ignition Delays

Abstract:

Currently, boron/barium chromate, also known as T-10, is predominantly used in the field of energetics as a pyrotechnic ignition delay. It has shown to be a reliable ignition delay in providing a consistent burning rates at varying effective diameters while also being a gasless reaction. However, barium chromate has been confirmed to be highly carcinogenic. This research focused on the additive manufacturing of five possible replacements for boron/barium chromate as a pyrotechnic ignition delay: zirconium/strontium molybdate, zirconium/barium molybdate, boron/strontium molybdate, boron/barium molybdate, and titanium/silicon. Each formulation utilized a low weight percent methyl cellulose solution in water as a printing medium to provide fluid-like behavior with high viscosity. Viscosity was used to characterize printing formulations, and performance behavior was based on burning rate vs. effective diameter and differential scanning calorimetry (DSC).

539118 The Role of Particle Size on the Combustion of Boron Carbide/Sodium Periodate Biocidal Formulations

Abstract:

The Role of Particle Size on the Combustion of Boron Carbide/Sodium Periodate Biocidal Formulations

Lance N. Kotter[a] and Lori J. Groven[b,*]

[a]South Dakota School of Mines and Technology, Department of Material Science and Engineering, Rapid City, SD, US

[b]South Dakota School of Mines and Technology, Department of Chemical and Biological Engineering, Rapid City, SD, US

***corresponding author email: lori.groven@sdsmt.edu**

Abstract

In our groups previous work, the boron carbide/iodate system has been shown to be an effective biocidal agent. The use of boron carbide mitigates aging issues typically associated with traditional aluminum-based formulations. In this work, boron carbide based biocidal formulations with $NaIO_4$ serving as the primary oxidizer are explored and the role of particle size on the combustion velocities, iodine output, and combustion products are detailed. Formulations are studied with and without the addition of binder and the role of particle size detailed toward developing 3-D printable energetics. It was shown in this study that the effect of changing the boron carbide particle size had negligible effects on both the combustion velocity and iodine recovery. Ignition was however significantly influenced. However, promising results have been established by reducing the particle size of the oxidizer, $NaIO_4$. The $NaIO_4$ was milled to an D_{90} value of 30.30μm resulting in iodine recovery close to 90%, much higher than any formulation examined thus far. By reducing the particle size even further to under 25 μm while maintaining the same fuel content, iodine recovery was close to 100%. Iodine quantification was done using UV-VIS, simultaneous differential scanning calorimetry and thermogravimetric analysis (DSC/TGA), heat of reaction, X-ray diffraction (XRD) and combustion characteristics will be discussed as a function of oxidizer size.

535864 Cermet-Based Hydrogen Transport Membrane Reactors for Conversion of Methane to Value-Added Chemicals

Abstract:

An abundance of natural gas presents new opportunities for use in the chemical industry, particularly gas-to-liquids (GTL) applications. The catalytic conversion of methane to benzene in the absence of oxygen is a reaction equilibrium-limited process (~12% at 700 °C, 1 atm) that may be enhanced by selective hydrogen removal. Cermet membranes consisting of a metal plated on a ceramic support provide the best combination of thermal stability, mechanical strength while enhancing the H2 flux. Nickel and palladium were electroless plated on porous alumina supports. The hydrogen separation performance of these membranes was evaluated for use in the methane-to-benzene process.

526882 Experimental and Numerical Study of an Intensified Water-Gas Shift (WGS) Reaction Process Using a Membrane Reactor (MR)/Adsorptive Reactor (AR) Sequence

Abstract:

Reactive separation processes, such as membrane reactors (MRs) and adsorptive reactors (ARs), have been attracting increased interest for industrial applications. Such a hybrid system combining a membrane reactor (MR) and an adsorptive reactor (AR) in tandem (with the AR following the MR, and the MRâ€™s reject stream serving as the ARâ €™s feed) can produce an ultra-pure H_2 product (without the need for using a post-processing step), and can deliver WGS reaction conversion exceeding equlibrium. The proposal configuration (AR following MR) provides significant flexibility in addition to high-purity hydrogen and carbon dioxide production.

The primary objective of this work is to first develop and simulate an advanced, and detailed mathematical model for the MR/AR sequence and then validate this model using experimental data obtained at the bench-scale. The MR system is composed of a reaction zone packed with catalyst pellets and glass beads, and a water/hydrogen permeation zone. In the AR, both catalyst and adsorbent materials are used for simultaneous reaction and separation. The AR is a dynamically operated process. Both steady-state and dynamic experiment are carried out in a laboratory setting to capture the individual MR/AR performances. Experiments are carried out for a broad range of temperatures (up to 300 ^{o}C) and pressures (up to 25 bar) relevant to realistic application conditions. The velocity, temperature and species concentration profiles in the MR reaction and permeation zones are captured by momentum/species transport models accounting for convection/reaction /diffusion/conduction mechanisms. The rigorous Maxwell-Stefan and dusty gas models are applied to describe mass diffusion fluxes. Finally, the performances of the combined system (the membrane reactor (MR) followed by an adsorptive reactor (AR) is simulated for a broad range of the operating conditions and design parameters. The simulation results are compared with the experimental findings.

532647 Hydrogen Production in Pd-Based Membrane Reactor Via Reforming Reactions

Abstract:

Abstract

In the last decades, the excessive use of fossil fuels to produce energy, as well as the growth of environmental pollution due to green house gases (GHGs) and other harmful emissions has led to the development of new technologies and to the utilization of renewable sources. In this context, hydrogen is recognized as one of the most promising energy carriers of the future. Unfortunately, hydrogen is not available in the Earth in elementary form and is currently produced primarily from steam reforming of natural gas, an energy-intensive process.

Nowadays, different methods, such as alcohols reforming reactions, and alternative devices, such as membrane reactors (MRs), are covering an important role for hydrogen production/separation. Specifically, alcohols as ethanol, methanol, glycerol etc. are largely available from renewable sources and they could be used as â€œbio-feedstocksâ€ for hydrogen generation via reforming reactions, while MR technology seems to represent a valid choice for replacing the conventional process due to its ability to provide the generation and separation of pure hydrogen in a single stage [1]. Among several renewable sources, bioethanol has received considerable attention due to its potential environmental benefits [2].

In the last years, Pd-based composite membranes, i.e. thin metallic layer supported on such porous materials as ceramics or stainless steel, have been considered owing to their lower cost (thin Pd layer) and higher mechanical resistant (porous support) than dense Pd-based ones [3]. Moreover, the use of Pd-alloys, in particular, with Ag, Au, Pt, is grown in the last years given their ability to enhance the hydrogen permeability and to resist sulfur poisoning.

Hence, the current study reviews the most relevant scientific results on hydrogen production from bio-ethanol reforming processes performed through membrane reactor technology. The benefits and the main challenges of inorganic membrane reactors, when applied to ethanol-reforming processes, are examined and the performances in terms of hydrogen yield, hydrogen recovery, and conversion are compared to those of conventional systems.

References

[1] A. Iulianelli, S. Liguori, A. Vita, C. Italiano, C. Fabiano, Y. Huang, A. Basile, The oncoming energy vector: hydrogen produced in Pd-composite membrane reactor via bioethanol reforming over Ni/CeO_2 catalyst, *Catal. Today*, 259 (2015) 368-375

[2] T. Yamazaki, et al., Behavior of steam reforming reaction for bio-ethanol over Pt/ZrO_2 catalysts, *Appl. Catal. B: Environ.* 99 (2010) 81-88.

[3] S. Liguori, et al., Performance of a Pd/PSS membrane reactor to produce high purity hydrogen via WGS reaction. *Catal. Today*, 193 (2012) 87-94.

531555 Na-LTA Membranes with High Water Selectivity for Dimethyl Ether Production in a Catalytic Membrane Reactor

Abstract:

As an important industrial intermediate, Dimethyl Ether (DME), can be used as a promising clean fuel in various ways. Traditionally, DME is produced either through methanol dehydration or a couple of consecutive reactions for DME directly from syngas. However, the yield of DME is low, limited by the equilibrium in both processes. In order to increase the DME yield, a novel catalytic membrane reactor to remove the produced water in-situ from the reaction system will be beneficial.

In this study, a highly water permeable NaA membrane, prepared on ceramic hollow fibers with secondary growth method, was investigated to selectively remove water from the simulated gas mixtures of $CO_2/H_2/CO$/Methanol/H_2O with representative compositions from the reactions producing DME from syngas route at very challenging conditions (temperatures up to 250 ^oC and pressures up to 600 psig). Current results showed that, when the feed water concentration was 2.57 mol.%, the membrane selectivity of H_2O/CO_2 was higher than 200, and no CO, H_2, Methanol were detected in the permeate side of the membrane at 250 ^oC and 300 psig. Stability testing over 11 h showed the NaA membrane had a good stability with <10% performance degradation. These results suggest that NaA membranes are capable of removing water in-situ in a membrane reactor, and thus increase the DME yield in syngas route at elevated temperatures and high pressures. We also expect NaA membranes be used to increase the product yield of reactions with water as a by-product in a catalytic membrane reactor at harsh conditions.

539731 PDMS/Ceramic Composite Membrane in Glycerol Fermentationâ€"PV Coupled Process for Biobutanol Production

Abstract:

Facing the depletion of fossil energy and the pollution of the environment, the development and utilization of biomass fuels have brought dawn to the sustainable development of human beings. Biomass fuel, as a renewable energy, has unparalleled advantages. With the development of biodiesel industry, the production of crude glycerol, as an inevitable by-product in the production of biodiesel, is huge. As we all know, bioconversion of crude glycerol without pretreatment to biofuels is the way to disposal of the surplus crude glycerol. In addition, biomass fuel butanol, which is produced by bio-fermentation, can be used as a clean renewable energy biofuel. At the same time, pervaporation is considered to be the greatest potential separation technology because of its energy-saving and efficiency, as well as no harmful effects on the microorganisms. Therefore, fermentation-PV coupled process is a membrane-based process so as to in-situ separate bio-product from fermentation broth by the alcohol-permselective membranes. PDMS membrane has been attracting more and more attention for the separation of biobutanol by coupling bio-fermentation process especially.

In this work, the PDMS/ceramic composite membrane separated the biobutanol directly from the crude glycerol fermentation broth, which not only relieves the inhibition effect of butanol, but also improves the yield of butanol. The pervaporation performance and the stability of the membrane in batch fermentation and fed-batch fermentation were systematically researched. During 250 h of continuous fed-batch fermentation-PV coupled process, the membrane exhibited high flux of 550 g/ (m^2 h) and separation factor of 24.1. Furthermore, compared with fermentation, the yield of butanol was increased by 55.6%. In this process, membrane fouling leaded to the degradation of membrane pervaporation performance after 200 h. The SEM, AFM, FT-IR and XPS characterizations demonstrated that the membrane fouling is due to the adsorption of active cells on the surface of PDMS membrane. Nevertheless, the total flux and separation factor of the water-washed membrane were the same as those of the fresh membrane, implying that the effect of membrane fouling on the membrane was reversible.

526892 Thermochemial Stability of ZIF Membranes for Membrane Reactor Applications

Abstract:

Thermochemical stability of the supported Zeolitic Imidazolate Framework (ZIF) membranes is studied for membrane reactor applications. It is found that the membrane/support interface acid-base chemistry has paramount inducing effects on the hydrothermal stability of ZIF membranes, resulting in the fatal acid-catalyzed decomposition of the ZIF frameworks. The ZIF-7/MgO-Al_2O_3 of marked thermochemical stability has permitted the first significant application of MOF membranes for catalytic membrane reactor in severe and practical process conditions, performing water-gas shift reaction ($CO + H_2O \leftrightarrow CO_2 + H_2$) at considerably high temperatures (473–673 K) and steam concentrations (20–40%).

517820 Enhancing CO_2/CH_4 Separation Performance and Mechanical Strength of Mixed-Matrix Membrane Via Combined Use of Graphene Oxide and ZIF-8

Abstract:

Gas separation using high-performance membranes have demonstrated itself as an energy-efficient and environment-benign process that is capable of replacing the conventional energy-intensive processes. In general, polymeric membranes have been widely incorporated in view of their good mechanical stability and processability. However, the trade-off effect between gas permeability and selectivity severely limits their future development of gas separation. Thus, mixed-matrix membranes (MMMs) that utilized synergistic effect of both polymer matrix and nanoporous fillers have been a focus of attention to researchers. This is because the addition of nanoporous materials acts as an efficient aid to favor selective gas transport. Nevertheless, the utilization of one type of filler tends to optimize either gas permeability or selectivity in gas mixtures. Therefore, in this context, an effective and feasible strategy in improving both gas permeability and selectivity was proposed by utilizing binary fillers.

High-performance MMMs that comprise both zeolitic imidazolate framework-8 (ZIF-8) and graphene oxide (GO) were synthesized to realize excellent CO_2/CH_4 separation. The incorporation of ZIF-8 nanocrystals alone in ODPA-TMPDA polyimide can be used to significantly enhance CO_2 permeability compared with that of pure ODPA-TMPDA. Meanwhile, the addition of a GO nanostack alone in ODPA-TMPDA contributes to improved CO_2/CH_4 selectivity. Hence, a composite membrane that contains both fillers displays significant enhancements in CO_2 permeability (up to 60%) and CO_2/CH_4 selectivity (up to 28%) compared with those of pure polymeric membrane. Furthermore, in contrast to the ZIF-8 mixed-matrix membrane, which showed decreased mechanical stability, it was found that the incorporation of GO could improve the mechanical strength of mixed-matrix membranes. Overall, the synergistic effects of the use of both fillers together are successfully demonstrated in this paper. Such significant improvements in the mixed-matrix membraneâ€™s CO_2/CH_4 separation performance and mechanical strength suggest a feasible and effective approach for potential biogas upgrading and natural gas purification.

540229 Accelerating Development and Intensification of Chemical Processes

Abstract:

Automated optimization and screening of production candidates along with process intensification offer opportunities for faster development and more efficient production for chemical industries. It is especially the case for those industries with many diverse products, such as specialty chemicals and pharmaceuticals. This presentation starts with opportunities in the early discovery/development phases enabled by automated screening and optimization of chemical reactions in microliter-scale droplets. The technology facilitates optimization across continuous variables (e.g., temperature, concentration, residence time) as well as discrete variables (e.g., catalyst species, base, and solvents) without requiring system depressurization or reconfiguration. Moreover, kinetic models can be extracted for subsequent scaling to process conditions.

As the next step in the development process, a robotically configurable chemical synthesis platform demonstrates automated, optimized multiple step continuous synthesis of molecular targets according to a predetermined recipe. Multistep synthesis occurs at elevated temperatures and pressures to enhance reaction rates, and the resulting residence times are on the order of minutes, in contrast to the multiple hour-long syntheses typically needed for batch processing. Process intensification is exemplified by on-demand synthesis of common pharmaceuticals, e.g., ciprofloxacin, in a plug-and-play, manually reconfigurable, refrigerator-sized manufacturing platform of integrated unit operations. CNC machining and 3D printing are used to realize the miniaturized process equipment described throughout the presentation.

The above advances in laboratory automation and translation to small-scale manufacturing promise to accelerate development, but planning chemical synthesis often requires considerable time and effort that can be reduced through computer-aided organic synthesis. Machine learning opens new approaches to retrosynthetic planning including the underlying challenges in reaction validation and forward reaction prediction. An overall workflow integrates the developed techniques for retrosynthesis and forward prediction to predict an ordered list of reaction paths to a given molecular target from purchasable starting materials via feasible reaction steps. The prospect for integrating these technologies into reaction engineering are discussed and shown through case studies.

530142 Particle Synthesis in Ultrasound-Integrated Microreactors

Abstract:

Microreactors and ultrasound are important process intensification tools with great potential especially for multiphase processes. As an example, the synergistic effect of coupling low frequency ultrasound with flow reactors to achieve cavitation have been proven successful in clogging prevention in solid-liquid flows.

In addition to their susceptibility for clogging in the presence of solid particles, microreactors are also characterized by their poor convective mixing, and the laminar velocity profile results in a wide residence time distribution which will also affect the size distribution of particles synthesized in them. To address these drawbacks, we focus on the design, modeling and experimental validation of a microreactor coupled with high frequency ultrasound. While high frequency ultrasound will not induce cavitation in the bulk liquid, the radiation force will affect particles present in the flow and deflect their trajectories. Furthermore, it can also excite secondary flow structures (known as acoustic streaming) which will improve the mixing in the bulk fluid.

In this work, we have designed an ultrasonic microreactor using a high frequency of 1 MHz, which ensures that the ultrasound is established as a standing wave in each reactor layer. Furthermore, the standing wave in the fluidic channel forms a nodal plane in its center, and the resultant acoustic radiation force will deflect all particle trajectories towards this nodal plane and thus focus the particles. This focusing effect prevents the wall deposition of synthesized particles, and in addition each particle will experience similar residence times leading to a reduced particle size distribution. The focusing efficiency of this ultrasonic microreactor is characterized using inert particles seeded with the flow. In a second step a solid forming reaction (precipitation of calcium carbonate) was performed. The effect of process parameters (e.g. ultrasound intensity, flow rate, particle concentration) on the focus extent and the particle size distribution (measured by laser diffraction) was investigated.

This work showcases how coupling high frequency ultrasound with microreactors can be exploited to improve particle synthesis.

534531 Demonstration of Scale-out Methodology for Intensified Liquid-Liquid Processes

Abstract:

DEMONSTRATION OF SCALE-OUT METHODOLOGY FOR INTENSIFIED LIQUID-LIQUID PROCESSES

*Eduardo Garciadiego Ortega, Dimitrios Tsaoulidis, Panagiota Angeli**

University College London, WC1E 7JE, London, UK

Abstract

Over the last few years there have been many studies on small scale units aiming at the intensification of multiphase processes relevant to energy and process industries. These include (bio)chemical synthesis (i.e. flow chemistry), absorption, extraction, and (nano)crystallisation. Gas-liquid and liquid-liquid processes carried out continuously in small channels exhibit fast mixing, high heat and mass transfer rates, and narrow residence time distributions when compared to conventional equipment. These advantages, in an industrial context, would translate to reduced costs, energy savings, diminished waste production, enhanced safety, and increased product quality.

The transition of these novel and efficient devices from bench-scale to the industrial application requires the throughput to increase without losing the merits of the small scales. The most common strategy to address this is by reproducing the small scale units in parallel in order to have the required throughput; this approach is termed scale-out or numbering-up. It contrasts the traditional scale-up where the size of one unit increases. Flow distributors are necessary to parallelise a process. Their effectiveness is critical to bridge the gap between bench and industrial scales for intensified processing. This challenge is accentuated in multiphase processes where maldistribution in either the residence time or the phase ratio in the main channels negatively affects the performance.

In this work double manifolds are considered for the two-phase flows distribution (Figure 1). They have several advantages, including, simple arrangement, small footprint and modularity. To model the flow distribution in the double manifolds, a new model has been developed following the resistance networks approach, similar to Al-Rawashdeh's model [1]. The model can be used to select the sizes in every section of the double manifold for a given maldistribution tolerance and required throughput.

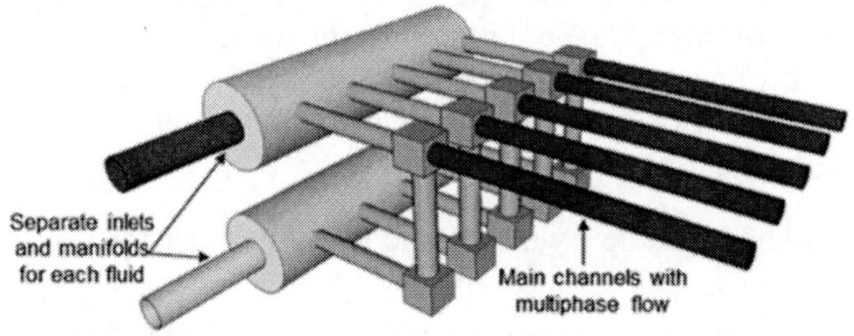

Figure 1. Schematic of a double manifold. Each fluid has a separate inlet; then both phases are distributed independently and mixed together into the main channels.

We present a demonstration of a step-by-step methodology to scale-out intensified liquid-liquid extractions. The process studied is the continuous extraction of U(VI) from a nitric acid solution into an organic phase using TBP as extractant. Previous to the scale-out, the single-channel process is characterised for conditions which result in plug flow. Hydrodynamic characteristics, such as features of the flow pattern (e.g. plug length and film thickness) and pressure drop, are related to the mass transfer performance. The results from the single-channel process are used to design a double manifold for the numbering-up. The resistance network model is used to decide the channel sizes in the manifold. Extraction studies are then carried out in the scale out system.

References

1. M. Al-Rawashdeh, F. Yu, T.A. Nijhuis, E. V. Rebrov, V. Hessel, J.C. Schouten, "Numbered-up gas-liquid micro/milli channels reactor with modular flow distributor," Chem. Eng. J. 207–208 (2012) 645–655. doi:10.1016/j.cej.2012.07.028.

537806 Multiphase Microchannel Separation Utilizing Capillary Pressure Gradients

Abstract:

A flexible, modular platform for multiphase separation is being developed for downstream processing that shows potential for significant reduction of capital cost and energy expenditure compared to conventional technology. The separator utilizes microchannel processing technology (MPT) to direct flow of each phase by creating a capillary force gradient via size and spacing of micro-scale architectural features, thereby controlling interfacial curvatures. With a proper selection of surface properties, the system is designed so that a selected phase cannot overcome capillary forces (in one direction of the gradient) with inertial and viscous forces of flow, guiding the fluid towards one outlet stream. The separator uses a flat â€œLamina Plateâ€ design that can accommodate a larger processing throughput per layer of device compared to single microchannel configurations. Microchannel-based separation presents significant gains in process intensification by accelerating mass transport and decreasing energy input per liquid volume compared to conventional methods. Beyond energy consumption, the scalable, distributed nature of MPT makes it amenable to distributed, smaller chemical manufacturing.

529629 Effect of Pressure on Ethane Dehydrogenation in MFI Zeolite Membrane Reactor

Abstract:

Using a membrane reactor (MR) for producing ethylene by ethane dehydrogenation (EDH) reaction is an effective way. Compared with packed bed reactors (PBR), the EDH MR effectively surpasses the equilibrium limit by timely removing H_2. A packed-bed membrane reactor (PBMR) with a Pt/Al_2O_3 catalyst was used to investigate the effect of pressure on EDH reaction. The EDH reaction was performed in the PBMR for the pressure and temperature range of 1-5 atm and 500-600 ^oC, respectively. With an increase in reaction temperature, the reaction rate increased which caused higher ethane conversion. Increasing reaction pressure helped in enhancing H_2 permeation across the membrane, which significantly increased the ethane conversion. The equilibrium limit of ethane conversion was successfully surpassed by increasing temperature and reaction pressure in the PBMR. Ethane conversion and ethylene selectivity as high as 29% and 97% were obtained at 600 ^oC and 5 atm for PBMR while corresponding values were 7% and 75% for PBR. The timely removal of H_2 from the reaction side also helped in reducing methane formation as H_2 is required for the methanation to occur. In addition, a 1D plug flow model was developed and the values for ethane conversion obtained from the model were validated with experimental results. The same model was used to evaluate the ethane conversion beyond the experimental conditions, showing ethane conversion >90% could be obtained.

537204 Scaling a Micro Structured Reactor for Sugar Chemistry from Lab Via Pilot to Full Production Scale

Abstract:

Scaling a micro structured reactor for sugar chemistry from lab via pilot to full production scale

M. Kraut[1)], G. Rabsch[1)] and R. Dittmeyer[1)]

1) Karlsruhe Institute of Technology, Institute of Micro Process Engineering, Hermann von Helmholtz Platz 1 76344 Eggenstein-Leopoldshafen, Germany

Motivation

It is well known, that micro structured devices show very high heat transfer rates, and some processes profit immensely by fast heating and cooling processes. Examples of production processes performed in micro structured reactors are still quiet rare. Following Roberge's work [1] it is surprising that not more processes have been transferred from macro scaled devices to micro or milli-scaled ones.

Especially processes in the food industry are often considered not to be compatible with micro structured flow devices due to the properties of natural products (varying quality, unknown side products, fouling tendency...) The challenge is taken up to make a new product using micro structured heat exchangers to obtain high specific heating rates to achieve defined residence times to perform a sugar transformation.

Process

The transformation of sugars at elevated temperatures is well known for a long time (e.g. caramel making). The target product is narrowly defined in terms of selectivity especially of undesired side products. This quality is achieved by a very well defined residence time of the reaction mixture at an equally well defined temperature. The reaction mixture firstly is brought to a temperature which is safely below the onset of both, desired and undesired reactions. Then the mixture is heated up rapidly (heating rates >100 K/s) to the optimum reaction temperature and held there for a defined time to obtain equilibrium yield of the desired and a minimum yield of undesired side product. After water is separated from the reaction mixture as steam the procedure is repeated with a second set of temperature and residence time.

Laboratory reactor

In Lab-scale the task was to find the parameters for operating the process reproducibly and safely within the required quality criteria. As a tool for preheating a symmetrical cubic heat exchanger was applied, which was described previously [2]. To perform the reaction and provide the necessary residence time a micro structured so called laboratory module[3] was used (fig 1, bottom).

This module provides 30 parallel reaction channels and for an intense heating 179 channels perpendicular to the reaction channels. The structured foil has been made from stainless steel (1.4571 or 316 Ti) and structured mechanically. The reactor is operated with a very high volume flow of thermos oil to obtain the boundary condition of a “hot wall”. Since the temperature difference in the oil is negligible a heat balance can’t be done, but the point is to provide the reaction with optimum conditions, not the most efficient heat usage.

Since the reaction itself has a very small enthalpy the heating effort is small, and one could argue that the reactor is over-designed. However, even small variation in temperature can lead to undesired effects, so that the condition to keep the temperature constant is sufficient to design a completely temperated reactor.

The process parameters for the reaction were found to be in the range of a few dozen seconds in residence time and a reaction temperature of about 200°C.

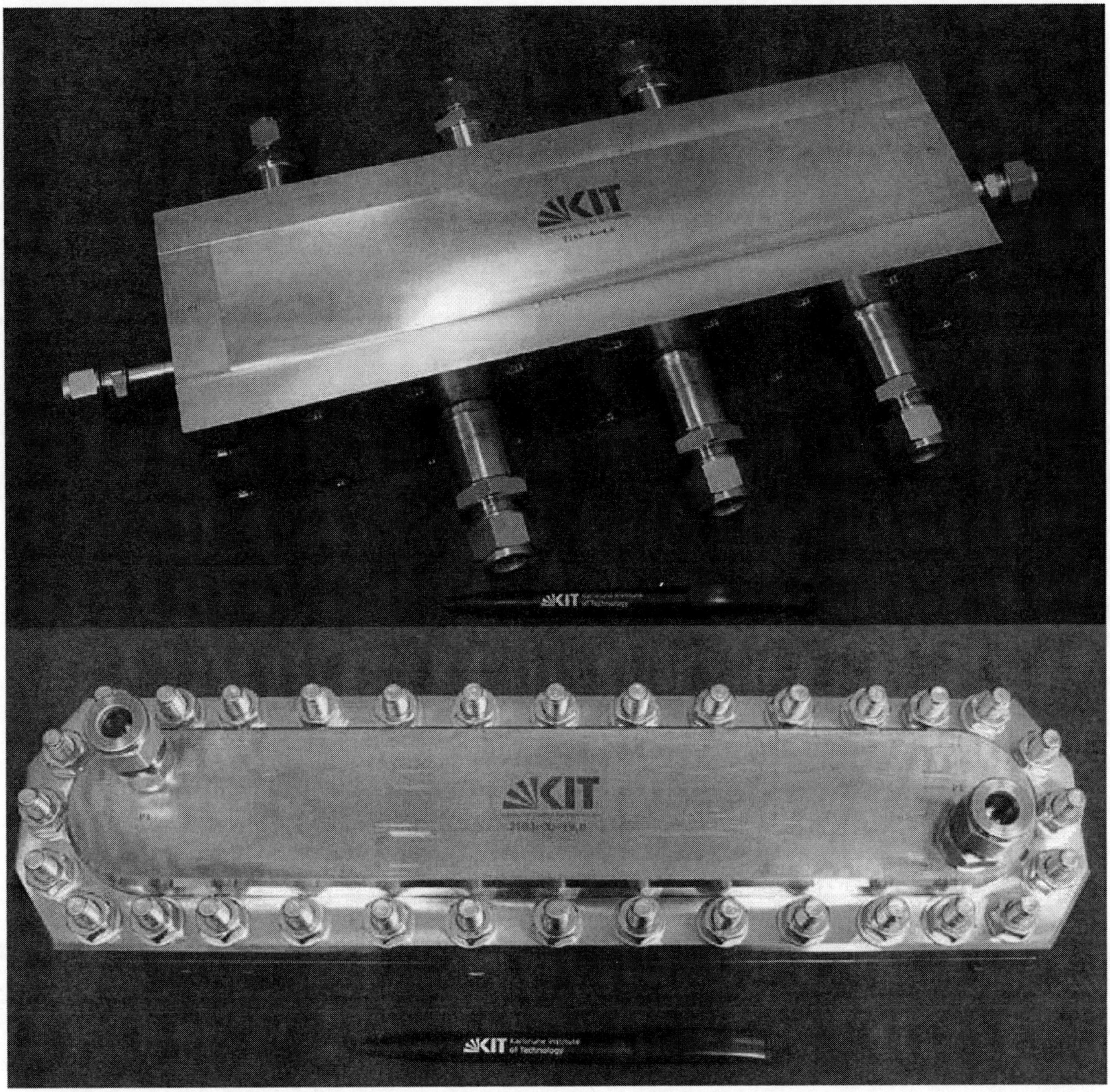

Fig.1 KIT lab module (bottom) and Pilot reactor derived from lab module (top)

Pilot scale reactors

The lab results were then transferred to a pilot scale. The scaling principle was to keep the structure of the device and multiply it according to the scaling factor, in this instance a factor of 15 (from 1 to 15 kg/h). Since the parameters of operation were known, the pilot device was manufactured in several layers of plates for heating and cooling respectively. The thickness of each foil was 1 mm, so the foils were diffusion welded to form a monolithic block (fig.2). The length of the heating channels was slightly increased to ensure diffusion welding compatibility of the device. However, geometry for the inflow has been altered, so that the specific pressure drop was not increased in comparison to the laboratory module. Also a 16th layer of heating channels was added to obtain the same conditions for each of the reaction channels (heated walls in both directions).

The reaction conditions were successfully reproduced at designed thoughput using this device. No detrimental scaling effects were observed. However, the flow distribution in the flanges was optimized during the pilot phase to ensure a better distribution based on cfd-calculations. [4]

To test the possibility of etching as an alternative manufacturing technique, which for higher amounts of foils is economically advantageous, one device was manufactured using etched foils. Dimensions were slightly altered to obtain circular channels by face-to face stacking. That device was also tested successfully.

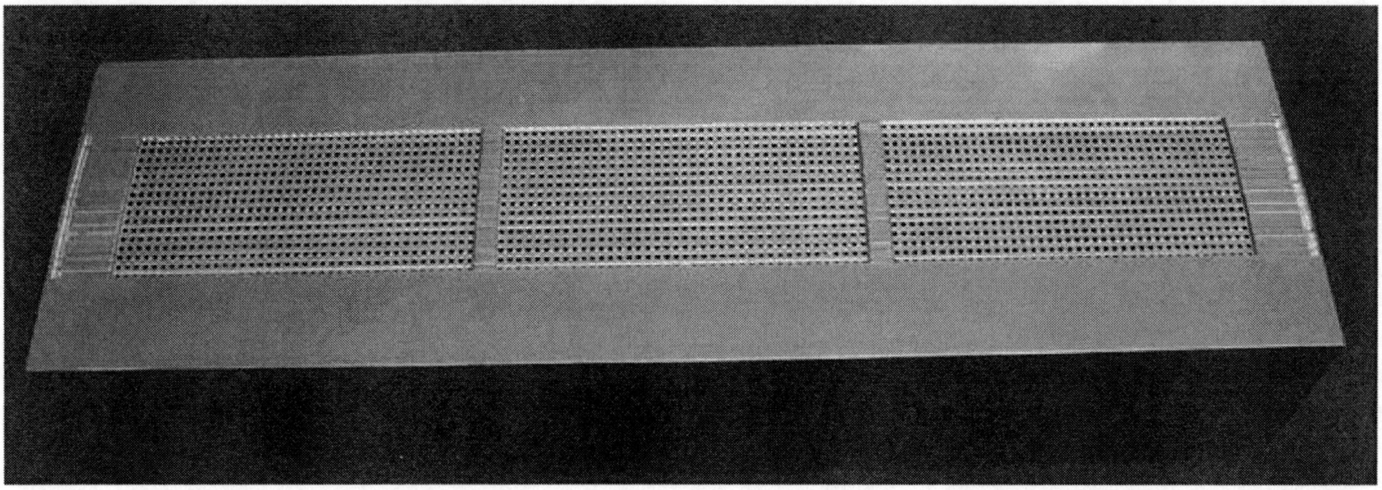

Fig.2 Diffusion welded block of pilot reactor. 16 layers of cooling foils are visible. Three cooling zones are evident.

Production reactor

Based on these encouraging results the design of a production unit with a further scaling factor of 60 was done. Since a reactor with a factor of 60 in height which would be the consequent scaling option was found to be not feasible; several conditions were altered:

- Crossflow heat exchange was replaced by countercurrent heat exchange
- The number of parallel reaction channels was increased
- The heating fluid is divided in two streams, each of which is flowing through a half of the reactor.

Design calculations ensure the condition of "hot wall" or constant reaction temperature along the whole channel.

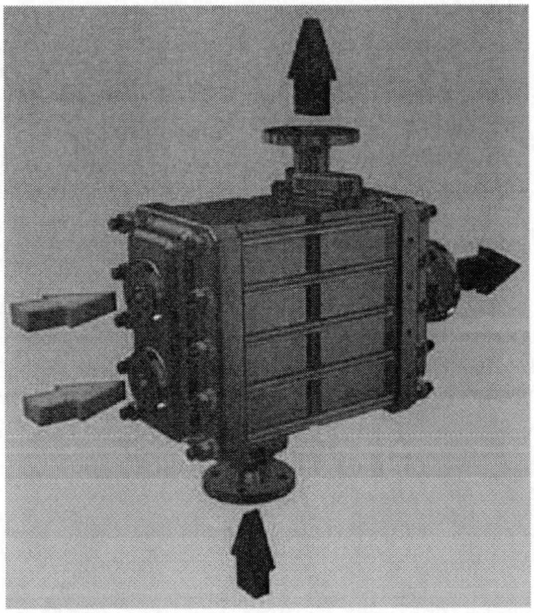

Fig.3: Design of production reactor as scaled up version of the pilot device, reaction direction ion green, heating in red.

Conclusions

A process of sugar transformation has been investigated using micro structured reactors and found to be feasible when using well defined residence time at equally well defined temperature, exploiting the high heating rates and flow conditions within micro structured channels.

A scale-up from laboratory device to a diffusion welded pilot device has been done successfully, showing that the scale-up strategy of using identical local conditions is feasible.

Further scale-up to obtain a micro structured reaction system has been performed with further economic improvements.

[1] Roberge, D.M., Ducry, L., Bieler, N., Cretton, P., Zimmermann, B., Chemical Engineering and Technology 28(2005), pp. 318-323

[2] Schubert, K., Brandner, J., Fichtner, M., (...), Schygulla, U., Wenka, A. , Microscale Thermophysical

Engineering5(2001), pp. 17-39

[3] K. Haas-Santo, M. Kraut, W. Benzinger, A. Wenka, L. Bohn, R. Dittmeyer, Chem. Eng. Tech., 82 (2010), p. 1341

[4] J. Cao, M. Kraut, R. Dittmeyer, L. Zhang, H. Xu, Int. Commun. Heat Mass 93 (2018), 60-65

521591 Green Synthesis of Polyvinyl Butyral (PVB) Using Microreactor System and Recycling Technology

Abstract:

We propose a highly efficient method for the synthesis of polyvinyl butyral (PVB) using microreactor system. The condensation reaction between polyvinyl alcohol (PVA) and n-butanal was successfully implemented, allowing the energy costs associated with cooling and mixing of the reactant solutions to be reduced. The microreactor system allowed the development of a recycling technology for the reuse of HCl, water, and excess n-butanal present in the product solution. In addition, with the aid of a two-step n-butanal feeding method, the developed recycling technology resulted in a 98.7% n-butanal utilization ratio, while the consumption of HCl and water during the reaction process was reduced by 85.6%. These results indicate that our novel process represents a more environmentally friendly approach to the PVB synthesis.

Funding support from National Key R&D Program of China (2017YFB0307102) and Tsinghua University Initiative Scientific Research Programs (20131089196, and 20151080361).

531887 A Reactive Distillation Optimization Model for Low- and High-Temperature Fischer–Tropsch Syntheses for Multiphase Product Recovery

Abstract:

Liquid fuels and chemicals are predominantly produced from crude oil. The abundance of shale gas has attracted attention for alternatives to be sought since natural or shale gas can be converted to the same liquid fuels and chemicals through integrated gas-to-liquid (GTL) processes. Biomass and carbonaceous wastes as well can also be converted to liquid fuels and chemicals through biomass-to-liquids and waste-to-liquids processes respectively. However, natural/shale gas is the most hydrogen-rich carbon source and can provide the most carbon-efficient route to fuels.

GTL processes require high capital cost but the main detractor of widespread adoption is investment risk linked to feed and products price volatilities. The decision to build a gas-to-liquid plant has to be based on the perceived future prices of oil and gas; and the prices of specialty chemicals produced from downstream processes. Therefore, robust technology that can offer product flexibility in response price volatilities is desirable to mitigate the investment risk.

Process intensification where syngas conversion in multiple reaction zones and product separation are combined in one unit can lead to smaller overall plants with lower capital and running costs. Consequently, we have developed an equation-oriented framework for optimal synthesis of integrated reactive distillation (RD) systems for FT processes in GAMS in which the phase equilibrium is given by a cubic equation of state, the reaction rate expressions are expressed in terms of fugacities, and the product selectivity of catalysts are based on the Anderson–Schulz–Flory distribution and experimental data. Comparing the RD for FT synthesis against conventional slurry reactors, provided results show that RD has a potential edge in industrial processes.

To explore the potential of producing a predominantly wax product on a low-temperature FT process, we are expanding the FT-RD optimization model to optimize tray numbers, feed tray location and catalyst loading on trays. We are also implementing the FT-RD optimization model within the PyOMO optimization framework. To produce a high gasoline fraction, light olefins, and some oxygenates (mainly alcohols) on a high-temperature FT process, we are developing a more detailed kinetic model to account for the various products.

536606 Design and Intensification of Sorption-Enhanced Reaction Processes for Methanol Production

Abstract:

Methanol is a viable alternate fuel to depleting oil and gas reserves, and is one of the most advantageous chemicals due to clean burning, CO_2 utilization and as a raw material for synthesizing hydrocarbons [1]. Since the commercialization of methanol production in 1923, the process has undergone significant improvements [2]. However, even after improved process efficiency, the current methanol production process from syngas is equilibrium-limited and is one of the largest energy consumers in chemical industry [3]. The process inefficiencies include low single-pass conversions, low methanol yield and selectivity, high energy consumption and equipment cost, and catalyst deactivation [4].

The objective of this work is to counter currently-existing process inefficiencies by enhancing single-pass reaction conversions and preventing catalyst deactivation. To this end, we apply the sorption enhanced reaction process (SERP) concept for designing a periodic sorption-enhanced methanol synthesis (SE-MeOH) process from syngas; *in-situ* removal of water byproduct favorably shifts the equilibrium towards enhanced methanol synthesis in accordance with the Le Chatelier's principle. For capturing the dynamics of the SE-MeOH process, we use the generalized adsorption-reaction modeling and simulation (GRAMS) platform. GRAMS is based on a one-dimensional, pseudo-homogeneous, non-isothermal, non-adiabatic, and non-isobaric NAPDE-based model, and has been extensively validated with experimental data for several SERP case studies [5]. GRAMS is coupled with an in-house simulation-based constrained grey-box optimizer for optimizing SE-MeOH process cycle configuration, design parameters and operating conditions [6][7]. We obtain single-pass carbon conversions as high as 80% in the Lurgi-type Methanol reactor via SE-MeOH with production capacity exceeding 109500 tons/yr, which is the typical capacity of an industrial plant [8]. The developed SE-MeOH process has smaller carbon footprint, enhanced product quality, and smaller reactor, condenser and recompressor size leading to significant savings in energy consumption and process economics. This presentation also covers the integration of the developed SE-MeOH process with a steam methane reforming reactor unit for converting natural gas to methanol via syngas route.

References

[1] G.A. Olah, Beyond oil and gas: the methanol economy, Angew. Chemie Int. Ed. 44 (2005) 2636–2639.

[2] A. Cybulski, Liquid-phase methanol synthesis: catalysts, mechanism, kinetics, chemical equilibria, vapor-liquid equilibria, and modeling—A Review, Catal. Rev. Eng. 36 (1994) 557–615.

[3] A. V Kruglov, Methanol synthesis in a simulated countercurrent moving-bed adsorptive catalytic reactor, Chem. Eng. Sci. 49 (1994) 4699–4716.

[4] M. Kuczynski, M.H. Oyevaar, R.T. Pieters, K.R. Westerterp, Methanol synthesis in a countercurrent gas—solid—solid trickle flow reactor. An experimental study, Chem. Eng. Sci. 42 (1987) 1887–1898.

[5] A. Arora, S.S. Iyer, M.M.F. Hasan, GRAMS: A General Framework Describing Adsorption, Reaction and Sorption-Enhanced Reaction Processes, Submitted. (2018).

[6] A. Arora, I. Bajaj, S.S. Iyer, M.M.F. Hasan, Optimal Synthesis of Periodic Sorption Enhanced Reaction Processes with Application to Hydrogen Production, Comput. Chem. Eng. 115 (2018) 89–111.

[7] I. Bajaj, S.S. Iyer, M.M.F. Hasan, A Trust Region-based Two Phase Algorithm for Constrained Black-box and Grey-box Optimization with Infeasible Initial Point, Comput. Chem. Eng. (2017).

[8] N. Rezaie, A. Jahanmiri, B. Moghtaderi, M.R. Rahimpour, A comparison of homogeneous and heterogeneous dynamic models for industrial methanol reactors in the presence of catalyst deactivation, Chem. Eng. Process. Process Intensif. 44 (2005) 911–921.

525385 Comparison of Adsorptive Separation Performance of a Hollow Fiber Bed and a Packed Bed: A Modeling Study Using Zeolite 13X and a Mixture of Propylene and Propane

Abstract:

Since its discovery by Carl Wilhelm Scheele in 1773, adsorption has become one of the commonly used industrial techniques for fluid (gas/liquid) separation, mostly fueled by the invention of tunable synthetic zeolites in the 1940s and the development of cyclic schemes which allowed for product recovery and adsorbent regeneration. [1] Traditionally, adsorbent crystals are formed into pellets (with or without a binder) and then packed into beds. However, pressure drops are usually incurred due to the random and tortuous nature of the packing void space which influences the cost of pressurizing the feed gas and the shape of the adsorption front. Heat transfer is inefficient for operating a TSA cycle and there is scope for improvement of fluid to adsorbent mass transfer coefficient. [2-4]

To overcome these difficulties, several structured configurations including, monoliths, laminates and foams, have been developed. Detailed reviews and comparisons of these configurations are available in literature [2,3,5,6]. More recently, a novel hollow fiber-based solid sorbent system has been proposed and experimentally validated[4,7-9] which provides an additional degree of control over the heat transfer efficiency as described here. The hollow fiber configuration relies on solid sorbents embedded in a porous polymeric hollow fiber matrix. Several identical fibers are assembled inside a module which resembles a shell and tube heat exchanger. The bore of the hollow fiber has an impermeable lumen layer which allows the flow of a cooling/heating medium to mitigate the temperature of operation of the process despite heat of adsorption/desorption. [4]

Compared to the traditional packed bed, the configuration of the hollow fiber bed offers several advantages: (1) the structured packing has a significantly lower pressure drop (2) the diffusion length for the fluid to the interior of the hollow fibers is significantly lower than the spherical packed bed (3) the bore allows for a heating/cooling medium to efficiently transfer heat to/from the adsorbent phase. However, in terms of adsorbent density per unit bed volume, the hollow fiber bed has a clear disadvantage.

From the previous discussion it is clear that while the use of a hollow fiber adsorbent bed has several advantages, it also has at least one significant drawback. It has been hypothesized that the advantages will outweigh the disadvantages of using a hollow fiber bed, but testing and understanding the drivers of the relative performance is needed. The aim of this computational study is therefore, to perform a detailed comparison of the properties and separation performance of the novel hollow fiber bed configuration with that of the traditionally used packed bed. To enable unbiased comparison, an equivalent model for the packed bed and the hollow fiber bed was developed. A preliminary parametric comparison of properties such as heat and mass transfer co-efficients, adsorbent density, and pressure drop was performed on the basis of these models. Optimized five step single bed cycles for the separation of propylene and propane on zeolite 13X were also studied. For the same recovery and purity, the hollow fiber bed was found to have a productivity that was five times higher than the packed bed. The hollow fiber bed also showed higher productivity, when parameters such as the desorption pressure and the ratio of the purge to feed velocity were varied in both beds.

1. DÄ…browski, A. Adsorptionâ€"from theory to practice. *Advances in colloid and interface science* **2001**, *93*, 135-224.
2. Rezaei, F.; Mosca, A.; Webley, P.; Hedlund, J.; Xiao, P. Comparison of traditional and structured adsorbents for CO2 separation by vacuum-swing adsorption. *Industrial & Engineering Chemistry Research* **2010**, *49*, 4832-4841.
3. Rezaei, F.; Webley, P. Optimum structured adsorbents for gas separation processes. *Chemical Engineering Science* **2009**, *64*, 5182-5191.
4. Lively, R. P.; Chance, R. R.; Kelley, B.; Deckman, H. W.; Drese, J. H.; Jones, C. W.; Koros, W. J. Hollow fiber adsorbents for CO2 removal from flue gas. *Industrial & Engineering Chemistry Research* **2009**, *48*, 7314-7324.
5. Rezaei, F.; Grahn, M. Thermal management of structured adsorbents in CO2 capture processes. *Industrial & Engineering Chemistry Research* **2012**, *51*, 4025-4034.
6. Akhtar, F.; Andersson, L.; Ogunwumi, S.; Hedin, N.; Bergström, L. Structuring adsorbents and catalysts by processing of porous powders. *Journal of the European Ceramic Society* **2014**, *34*, 1643-1666.
7. Lively, R.; Chance, R. R.; Koros, W. J.; Deckman, H. W.; Kelley, B. T.: Sorbent fiber compositions and methods of temperature swing adsorption. Google Patents, 2008.
8. Fan, Y.; Labreche, Y.; Lively, R. P.; Jones, C. W.; Koros, W. J. Dynamic CO2 adsorption performance of internally cooled silicaâ€supported poly (ethylenimine) hollow fiber sorbents. *AIChE Journal* **2014**, *60*, 3878-3887.
9. DeWitt, S. J.; Sinha, A.; Kalyanaraman, J.; Zhang, F.; Realff, M. J.; Lively, R. P. Critical Comparison of Structured Contactors for Adsorption-Based Gas Separations. *Annual review of chemical and biomolecular engineering* **2018**.

526769 Towards a Systematic Process Intensification Framework for Advanced Distillation Systems

Abstract:

Towards a Systematic Process Intensification Framework for Advanced Distillation Systems

Yuhe Tian[a,b,c], M. Sam Mannan[a,b,c], and Efstratios N. Pistikopoulos[a,b]

[a] Artie McFerrin Department of Chemical Engineering, Texas A&M University, College Station, TX

[b] Texas A&M Energy Institute, Texas A&M University, College Station, TX

[c] Mary Kay O'Connor Process Safety Center, Department of Chemical Engineering, Texas A&M University, College Station, TX

Process Intensification (PI) offers the potential to drastically reduce the energy consumption and processing cost by utilizing multifunctional phenomena at different time and spatial scales. Distillation, the most widely utilized but highly energy-intensive separation technology, accounts for over 40% of the energy used in the chemical process industry and over 95% of that in separation industry [1]. Various advanced distillation technologies (e.g., heat-integrated distillation, reactive distillation, extractive distillation, and membrane distillation) have been developed based on PI principles to maximize synergy effects and can achieve around 30% energy savings [2]. Despite recent advances [3-6], systematic synthesis methods and software prototypes to explore phenomena-based process optimization and intensification for advanced distillation design are still lacking.

To address this challenge, in this work, we propose a systematic process intensification and synthesis framework based on recent extensions of the phenomenological Generalized Modular Representation Framework [7-9]. Herein, distillation systems are represented as aggregated multifunctional mass/heat exchange modules based on which intensification possibilities can be revealed without a pre-postulation of equipment or flowsheet configurations. Process driving forces, described by Gibbs free energy changes, are optimized to enhance fundamental transport and reaction performances. The derived intensified designs can then be analyzed by detailed steady-state simulation and design optimization. The capabilities of the proposed framework to deliver advanced distillation systems with improved energy efficiency are demonstrated with three case studies: (i) heat-integrated distillation, (ii) extractive distillation, and (iii) reactive distillation. Extensions of the framework to include process safety and dynamic operability assessment will also be discussed [10].

Reference

1. Kiss, A. A. (2014). Distillation technology–still young and full of breakthrough opportunities. *Journal of Chemical Technology and Biotechnology*, *89*(4), 479-498.

2. Olujić, Ž., Jödecke, M., Shilkin, A., Schuch, G., & Kaibel, B. (2009). Equipment improvement trends in distillation. *Chemical Engineering and Processing: Process Intensification*, *48*(6), 1089-1104.

3. Tian, Y., Demirel, S. E., Hasan, M. M. F., & Pistikopoulos, E. N. An Overview of Process Systems Engineering Approaches for Process Intensification: State of the Art. *Chemical Engineering and Processing: Process Intensification*, Submitted.

4. Tula, A. K., Babi, D. K., Bottlaender, J., Eden, M. R., & Gani, R. (2017). A computer-aided software-tool for sustainable process synthesis-intensification. *Computers & Chemical Engineering, 105*, 74-95.

5. Pichardo, P., & Manousiouthakis, V. I. (2017). Infinite DimEnsionAl State-space as a systematic process intensification tool: Energetic intensification of hydrogen production. *Chemical Engineering Research and Design*, *120*, 372-395.

6. Demirel, S. E., Li, J., & Hasan, M. F. (2017). Systematic process intensification using building blocks. *Computers & Chemical Engineering, 105*, 2-38.

7. Papalexandri, K. P., & Pistikopoulos, E. N. (1996). Generalized modular representation framework for process synthesis. *AIChE Journal*, *42*(4), 1010-1032.

8. Ismail, S. R., Proios, P. and Pistikopoulos, E. N. 2001. Modular synthesis framework for combined separation/reaction systems. *AIChE Journal*, *47*(3), 629-649.

9. Proios, P., Goula, N. F., & Pistikopoulos, E. N. (2005). Generalized modular framework for the synthesis of heat integrated distillation column sequences. *Chemical engineering science*, *60*(17), 4678-4701.

10. SYNOPSIS – Synthesis of Operable Process Intensification Systems. Awarded project [DOE RAPID Manufacturing Institute].

535419 An MILP-Based Operability Approach for Process Design, Intensification and Modularity of Nonlinear and High-Dimensional Energy Systems

Abstract:

Operability emerged as a tool for the selection of the necessary inputs to achieve specified desired outputs in a given process [1]. In a previously developed mixed-integer linear programming (MILP)-based operability approach [2], the operability concept was extended to attain process intensification towards obtaining modular designs for energy systems. While a computational time improvement was achieved, the developed framework is still restricted to computational geometry techniques applied in two-dimensional spaces. The extension of this approach to systems with input and output spaces of higher dimensions consists of a challenge that still remains to be addressed. High dimensions give rise to new concerns such as volume of simulated data and overall increase in algorithm computational complexity, potentially impacting the computational cost. In this work, an improved MILP-based operability framework is derived with the utilization of computational geometry techniques suitable for higher-dimensional and nonlinear systems. This presentation will be focused on the discussion of the feasibility of the new approach to produce intensified modular designs for energy systems applications.

Process operability has been successfully applied to high-dimensional and nonlinear systems by the utilization of nonlinear programming (NLP) tools [3]. Potential tractability issues associated with the high computational expense of nonlinear algorithms have been tackled with the employment of parallel computing [4]. In the proposed framework, the addressed high-dimensional and nonlinear system is decomposed into several linearized subsystems. These subsystems are obtained using spatial discretization techniques that produce a set of convex polytopes in each n-dimensional space. Then, an MILP optimization problem is formulated based on spatial search and barycentric interpolation. For this application, the goal of the optimization problem is to minimize physical size, satisfying all process constraints and intensification targets. The outcome of this optimization problem is a modular design of high efficiency, associated with the most intensified operating point. As MILP tools have higher computational efficiency than NLP tools, a single processor will be considered for the proposed framework calculations, in contrast to the parallel workers present in the NLP approach.

The developed framework will be tested and compared to past NLP approaches for two case-studies: a membrane reactor for the direct methane aromatization conversion to hydrogen and benzene and a natural gas combined cycle (NGCC) process. These systems are characterized by dimensionalities as high as 5x5 and 8x8 dimensions, respectively [4]. The trade-off between accuracy and computational cost will be discussed for the validation of the proposed framework when compared to NLP approaches for high-dimensional system extrapolations.

References

1. Vinson D. R. and Georgakis C. "A new measure of process output controllability". J. Proc. Cont., 10(2-3), 185-194 (2000).
2. Gazzaneo V., Carrasco J.C. and Lima F. V. "An MILP-based operability approach for process intensification and design of modular energy systems." Accepted for publication in Proceedings of Process Systems Engineering (PSE), July (2018).
3. Carrasco J.C. and Lima F. V. "Operability-based approach for process design, intensification, and control: application to high-dimensional and nonlinear membrane reactors". In Proceedings of the FOCAPO/CPC (2017).
4. Carrasco J.C. and Lima F. V. "Bilevel and parallel programming-based operability approaches for process intensification and modularity." Accepted for publication in AIChE Journal. DOI10.1002/aic.16113 (2018).

525666 Compact, Nonsmooth Operators for Single-Component Mass and Water Integration

Abstract:

Focus in modern plant retrofit, expansion, and design has shifted towards the use of intensification techniques to implement processes that are both sustainable and economically competitive. By calculating the minimum resource usage required for a process with optimal reuse, process integration methods can be used to identify targets where there is significant scope for intensification and motivate the design of novel technologies, such as multistream heat exchangers for optimal heat recovery. Therefore, process integration can be used both to intensify existing plants where a fundamental redesign of the process is infeasible and also to ensure that proposed intensified processes operate optimally. To further decrease resource use, these integration methods have expanded beyond integrating individual processes to sharing resources between plants in the form of eco-industrial parks. However, as process integration is applied to increasingly complex problems, improved numerical methods are required to efficiently consider resource usage and constraints.

Current methods for solving process integration problems fall into two general categories: pinch analysis or superstructure approaches. Pinch analysis methods are physically intuitive and relatively easy to implement; however, current techniques are often heuristic and limited to a small class of problems. Even algorithmic pinch point approaches are typically mathematical programming problems with numbers of constraints that increase significantly with the number of sources and sinks. In the context of fixed-flow rate water allocation problems, these approaches are also unable to solve for concentration process variables and cannot be used for simultaneous process integration and optimization. On the other hand, superstructure techniques are adaptable to many different integration problems. However, these formulations are mixed-integer programming problems that often involve nonconvex functions, are difficult to solve, and whose complexity increases rapidly with problem size.

To address these limitations, we have developed compact, nonsmooth operators for both single-contaminant mass and water integration problems. These operators extend the work of Watson *et al.*, who simulated multistream heat exchangers using a nonsmooth system of equations adapted from a mathematical programming pinch analysis approach proposed by Duran and Grossmann [1,2]. Our formulation consists of systems of only two equations: an overall resource balance for the process and a single nonsmooth equation describing the resource balance below potential pinch points. Automatic lexicographic-directional differentiation is used to calculate exact generalized derivative elements for the nonsmooth equation system [3]. These elements are used in a suitable method, such as Semismooth or Linear Programming Newton, to solve the system without requiring smoothing approximations or additional variables.

We will present these novel operators and their solution methods and demonstrate how they can be used both to solve for optimal resource targets and for process parameters given resource constraints, including source and sink concentrations in water allocation problems. In addition, we will show other unique benefits of this approach, including the ability to simulate both pinch and threshold problems using a single model. Overall, our results will highlight how these nonsmooth operators significantly reduce problem complexity compared to optimization approaches and can provide computationally tractable solutions to large-scale interplant integration problems.

References

[1] H .A. J. Watson, K. A. Khan, and P.I. Barton. Multistream Heat Exchanger Modeling and Design. AIChE Journal, 61(10):3390-3403, 2015.

[2] M.A. Duran and I.E. Grossmann. Simultaneous Optimization and Heat Integration of Chemical Processes. AIChE Journal, 32(1):123-138, 1986.

[3] K. A. Khan and P. I. Barton. A vector forward mode of automatic differentiation for generalized derivative evaluation. Optimization Methods and Software, 30(6):1185-1212, 2015.

535724 Optimal Mass Exchanger Network Synthesis Using a 2-Step Hybrid Algorithm Including Packed Column Design

Abstract:

A method for the synthesis of mass exchanger networks (MENs) is presented that focuses on including detailed packed column design considerations at the network optimization level to deliver designs that are optimal and more realistic. Most approaches to simultaneous optimization of MENs have been based upon a stage-wise superstructure with simplified representations of the individual mass exchangers. Even with these simplified representations, the MINLP is still non-convex and computationally difficult to solve with modern deterministic solvers. Additionally, it is not possible to guarantee that when a solution is found that it is globally optimal.

The method developed in this study, uses a version of this aforementioned stage-wise MINLP formulation in the first step. Once the solution to this is found, a secondary nonlinear programming (NLP) step, based on orthogonal collocation on finite elements, is performed to find the optimal column dimensions, packing sizes, fluid velocities, and pressure drops, based on the flows and stream matches from the network optimization. Once the new individual column designs are found, the MINLP is then updated with a series of correction factors based on the solution of the detailed designs. This procedure repeats between the MINLP and NLP sub-optimization until the differences between the solutions converge or until a maximum number of iterations is reached.

The methodology, presented in detail in the work of Short, et al. (2018), is moved from the GAMS environment into PYOMO, allowing for powerful new ways to enhance the algorithm. The previous methodologyâ€™s correction factor determination is improved upon in order to speed up and increase the chances of convergence. Additionally, through the application of different MINLP solvers, larger examples that were previously not able to be solved are included.

Reference: Short, M., Isafiade, A.J., Biegler, L.T.,Kravanja, Z., 2018, Synthesis of mass exchanger networks in a two-step hybrid optimization strategy, Chemical Engineering Science, 178, 118 - 115.

529420 Mathematical Optimization of Sustainable Water Desalination Processes Using Directional Solvent Extraction

Abstract:

Access to clean fresh water is an ever-growing concern for modern society as it is critical to ensure human health, to protect threatened ecosystems, and to promote economic growth and prosperity. Modern seawater desalination technologies remain energy intensive; they required three to four times the theoretical minimum energy for separation [1]. Further research is required to both reduce the energy intensity of seawater desalination and design new systems that are driven by renewable solar thermal and waste process heat.

Directional solvent extraction (DSE) is a promising new technology for solar desalination that can effectively utilize low temperature (40 °C to 80 °C) heat. DSE exploits special solubilities of certain solvents (typically saturated fatty acids) to extract fresh water from saline sources at elevated temperature and release the water when the solution is cooled. This approach has several unique features: (1) water can dissolve in the solvent and the solubility increases with temperature; (2) the solvent is virtually insoluble in water; (3) the solvent does not dissolve salts; and 4) no membrane is required. Previous work includes initial concept demonstration as a batch process [2,3], molecular simulation to understand solvent performance [4], and heat integration for a single-stage continuous process [5]. The predicted energy requirement for DSE, without process intensification or optimization, is >160 kWh/m^3 of fresh water.

This presentation explores new process intensification opportunities for DSE. Most notably, previous work has focused on single-stage configurations [5]. Preliminary results suggest a counter-current configuration with a moderate number of equilibrium stages can reduce the energy requirement by up to 70%. Moreover, a mathematical optimization framework is proposed to optimize the DSE process and systematically explore trade-offs between product quality, energy requirements, and capital costs. This is done using an equation-oriented approach that simultaneously optimizes process operating conditions (such as temperatures, flow rates, compositions), design parameters (such as equipment sizes), and heat recovery opportunities [6,7]. Thus, the simultaneous approach can exploit tens to hundreds more degrees of freedom than conventional targeting methods (i.e., as employed in [5]), which consider optimal heat integration with fixed flowrates and temperatures. This capability is critically to fully optimize the DSE process and inform ratio molecular design of novel thermoresponsive solvents.

References:

[1] M. Elimelech, W. A. Philip (2011). The Future of Seawater Desalination: Energy, Technology, and the Environment. *Science* 333 (6043), pp. 712-717.

[2] A. Bajpayee, T. Luo, A. Muto, G. Chen (2011). Very low temperature membrane-free desalination by directional solvent extraction. *Energy Environ. Sci.* 4, pp. 1672â€“1675.

[3] D. Rish, S. Luo, B. Kurtz, T. Luo (2014). Exceptional ion rejection ability of directional solvent for non-membrane desalination. *Appl. Phys. Lett.* 104 (2), p. 024102.

[4] T. Luo, A. Bajpayee, G. Chen (2011). Directional solvent for membrane-free water desalinationâ€”A molecular level study. *J Appl. Physics.*, 110 (5), p. 054905.

[5] S. Alotaibi, O. M. Ibrahim, S. Luo, T. Luo (2017). Modeling of a continuous water desalination process using directional solvent extraction. *Desalination* 420, pp. 114-124.

[6] A.W. Dowling, L. T. Biegler (2015). A Framework for Efficient Large Scale Equation-Oriented Flowsheet Optimization. *Comp. Chem. Eng.* 72, pp. 3-20.

[7] A. W. Dowling, L. T. Biegler (2013). Optimization-based process synthesis for sustainable power generation. *Chem. Eng. Trans.* 35, pp. 1-12.

537967 Photovoltaic/Thermal Energy Addition to Expeditionary Buildings

Abstract:

The US Air Force (USAF) has a goal of reducing field consumption of fossils fuels. One cost-effective solution is to retrofit existing field assets with the ability to harness renewable energy. This paper discusses the adaptation of USAF expeditionary shelters with photovoltaic thermal (PVT) hybrid panels. The resulting power is captured as both electrical energy and thermal energy.

An Air Force Civil Engineering team constructed steel buildings (known as K-Spans) at Tyndall Air Force base (Florida) as test structures for evaluation of the PVT hybrid technology, which is attached to the outside.

The capabilities, operating characteristics, and overall power gain of the configuration were examined, and results will be presented. Varying flowrates were used to investigate the trade-off between heated water and higher electrical efficiency. Similarly, control algorithms were used to control the flow and output water temperatures compared to a non-controlled baseline unit.

522844 An Integrated Approach to Water-Energy Nexus with Multiple Energy Sources

Abstract:

AN INTEGRATED APPROACH TO WATER-ENERGY NEXUS WITH MULTIPLE ENERGY SOURCES

Fadhil Y. Al-Aboosi and Mahmoud M. El-Halwagi

The Artie McFerrin Department of Chemical Engineering

Texas A&M University

College Station, Texas 77843-3122

ABSTRACT

Growing energy demand, water scarcity, and climate change have emerged as considerable challenges to sustainable development. Consequently, there is a necessity to create the proper management strategies for water resources (e.g., wastewater treatment) and to integrate conventional energy sources with renewables (e.g., solar energy, wind energy, biofuels, etc.). The objective of this study is to develop a design framework for integrating water and energy systems including multiple energy sources, cogeneration process, and desalination technologies. Solar energy is included to provide thermal power directly to a multi-effect distillation plant (MED) exclusively (to be more feasible economically) or indirect supply through a thermal energy storage system. Thus, MED is driven by direct or indirect solar energy, and excess or direct cogeneration process heat. The proposed thermal energy storage along with the fossil fuel boiler will allow for the dual-purpose system to operate at steady-state by managing the dynamic variability of solar energy. Additionally, electric production is considered to supply a reverse osmosis plant (RO) without connecting to the local electric grid. A multi-period mixed integer nonlinear program (MINLP) is developed and applied to discretize operation period to track the diurnal fluctuations of solar energy. The solution of the optimization program determines the optimal mix of solar energy, thermal storage, and fossil fuel to attain the maximum annual profit of the entire system. A case study is solved for water treatment and energy management for Eagle Ford Basin in Texas.

Keywords: Cogeneration, Process integration, Solar energy, Thermal storage, Desalination, Optimization

538748 The "Green Latrine": Deployable Latrine with Off-Grid Photovoltaic/Thermal Energy

Abstract:

The US Air Force has a goal of reducing field consumption of fossils fuels in order to improve energy efficiency and operational resiliency. One cost-effective solution is to provide field assets that can operate independent of the power grid by harnessing renewable energy. This paper details the design, construction, and assessment of a proof of concept deployable latrine powered exclusively with Photovoltaic/Thermal (PVT) energy technology that allows operation completely "off-grid". PVT arrays are traditionally used for parallel production of solar electrical power and water heating in residential recreational pools. The attachment of a heat exchanger to the back of a traditional PV panel allows for cooling of the panels, resulting in higher electrical output while simultaneously heating the water.

A standard military portable (containerized) latrine unit was retrofitted with PVT hybrid panels that generate hot water and electricity. Potable water is the only required input. The installed electrical production capacity totaled 2.475 kW with an associated battery bank (44 kWh). The PVT added 6.0 kW of additional potential thermal energy depending upon variation in the panel flow rates and PVT configuration.

The capabilities, operating characteristics, and overall power gain of the configuration was examined during studies at Tyndall Air Force Base, Florida, and results will be presented. The system's performance is examined under varying operating scenarios to assess electrical and thermal performance. Overall energy efficiency was determined in respect to electrical capacity, electrical storage, and operating utility.

539205 Optimization of Hybrid RO-PRO Membrane Processes at the Water-Energy Nexus

Abstract:

Water and energy are two of the most pressing problems faced by the world today. Membrane processes such as Reverse Osmosis (RO) and Pressure Retarded Osmosis (PRO) can help alleviate both challenges. This work focuses on systematic optimization of a two-stage hybrid RO-PRO process for energy-efficient seawater desalination which integrates RO for desalination and PRO for power generation into a single process. Such a process is of great significance at the water-energy nexus given that water production, pump energy consumption, and hydraulic/osmotic energy recovery are inherently intertwined. The optimization problem is formulated to minimize the Specific Energy Consumption (SEC) subject to specified total membrane area and total water recovery. By solving the optimization model the best design (e.g., allocation of membrane area among different units) and operating (e.g. applied pressure in each membrane unit) conditions are simultaneously determined. It is shown that the proposed two-stage RO-PRO is energy-efficient than other configurations currently available in literature for continuous membrane-based RO desalination. Challenges for commercial viability of RO-PRO are discussed.

Reference:

M. Li. Optimization of Hybrid RO-PRO Membrane Processes at the Water-Energy Nexus, under review, 2018.

520120 An Integrated and Distributed Anaerobic Digestion Waste-to-Energy System for Energy Recovery from Food Waste

Abstract:

High energy self-consumption and low energy production are two major reasons limiting the energy performance of waste-to-energy systems. As a major part of energy consumption, mixing in an anaerobic digester creates a homogeneous system enhancing mass transfer, and enables the solid wastes and microorganisms remain in suspension, but continuous mixing strategy is not cost-effective due to the demand of high electric energy. A high-energy efficient food waste-to-energy system was developed to combine anaerobic digestion and engine generator for energy recovery from food waste. Through investigating the effect of different mixing strategies on the performance of the anaerobic digestion process, mixing time was optimized to lower energy consumption and ensure that biogas production is economical and energy efficient. By testing an internal combustion engine generator, electricity generation potential and waste heat behavior were obtained when using different components of methane gas as fuel. Raising the methane content over 85% enhanced the performance of the engine generator, while the methane content at 60% was unable to drive the engine stably. In addition, the waste heatâ€™s utilization and electricity generation efficiency, as well as energy distribution were studied through the analysis of the whole system.

537732 Chemical Looping Based Hydrogen Production from Ammonia: System Analysis and Experiments of a Two Reactor and a Three Reactor System

Abstract:

Hydrogen (H_2) is a promising carbon neutral liquid fuel with a high specific energy of 39.8 kWh/kg (HHV). The market penetration of H_2 is limited by the high transportation costs associated with its low energy density of 0.8kWh/L (by comparison gasoline is 8.76 kWh/L) . A potential solution to reduce transportation costs for H_2 is to use ammonia (energy density: 4.25 kW/L). Ammonia (NH_3) is proposed to be an alternative H_2 carrier, due to its lower transportation costs and it is widely produced on an industrial scale using Haber-Bosch chemistry. From an end-user perspective, if ammonia is used as a H2 carrier and if H2 is still used as the primary energy source, there exists a technological need to convert NH3 back to H2 in an efficient way.

OSU has proposed a thermochemical solids looping based based NH3 to H2 (ATH) process that can operate at high thermal and H2 production efficiencies. The ATH system can be operated as a two reactor system or a three reactor system, wherein the first the reducer reactor utilizes an intrinsic O2 gradient driven by the reduction potential of NH3 and a metal oxide Fe2O3/Fe3O4) to efficiently crack NH3 to a mixture of N2, H2 and H2O. The reduced metal-oxide from the reducer reactor is re-oxidized in a second reactor oxidizer using H2O as the oxygen source.

This presentation will initially compare and contrast the two reactor ATH system to the three reactor ATH system. Specifics of the thermodynamic rationale for optimizing the H2 production from NH3 using these configurations, while minimizing the operating temperature will be presented. The study uses guidance from the thermodynamic simulations to design reaction-engineering experiments in a thermo-gravimetric analyzer and a fixed bed system for generating proof of concept data. A techno-economic analysis model based on reaction engineering parameters is constructed and analyzed using sensitivity studies that provide insights into further process development will also be presented.

540933 Process Intensification of Bio-Ethanol Dehydration Under Compressed Liquid Phase Conditions

Abstract:

Global ethylene demand is consistently growing as its uses as a precursor for plastics, detergents and lubricants expand. Although the majority of ethylene is produced from petrochemical sources, an existing green alternative is dehydration of ethanol derived from biorenewable feeds to produce ethylene. High feed costs and process inefficiencies limit bio-based ethanol as an economic alternative to crude oil for ethylene production. Specifically, the highly endothermic bioethanol dehydration reaction and poor heat transfer to the vapor phase necessitate operation through a series of adiabatic reactors and boilers to maintain high ethylene selectivity. In addition, agriculture and fermentation are highly localized, which necessitates costly ethanol transportation through trucking, waterways, and pipelines. Operation of ethanol dehydration in the compressed liquid phase has potential to intensify the dehydration process by improving heat transfer, product recovery, and energy efficiency. A process intensified reactor operating under liquid phase conditions has potential to reduce capital costs sufficiently to justify distributed ethylene production.

Catalytic activity in the liquid phase is a major knowledge gap that hampers development of process intensified, liquid phase ethanol dehydration reactors. To assess the effects of conditions on activity, we performed both liquid and vapor phase ethanol dehydration studies, studying both space velocity and the effect of the water content of the fed. Ethanol dehydration reactions were performed in a continuous phase packed bed reactor catalyzed using ZSM-5 zeolite. Ethylene activity was compared when operated under both liquid and vapor phase (0.1 or 24 MPa) and varying weight hourly space velocities (30 - 1200 hr$^{\text{â€˜1}}$) and water loadings (0 - 33 wt%). ZSM-5 is more active for ethanol dehydration reactions under vapor phase conditions compared to liquid phase conditions, however, differences in ethylene yield are negligible when operated at low space velocities and high water loadings. Characterization of ZSM-5 catalyst after reaction reveals only a minor loss in framework after both the vapor and liquid phase reaction. Ethanol dehydration throughput and heat transfer improvements using a compressed, liquid phase packed bed reactor compared to vapor phase reactors suggest a potential alternative to current industrial processes.

538702 Real Time Constrained Model Predictive Control for the Continuous Manufacturing of Pharmaceuticals

Abstract:

The cost of healthcare in the US is the fastest growing component of the inflation index. One method aimed at reducing these growing costs is in the development of processes for continuous pharmaceutical manufacturing. The continuous production of pharmaceuticals offers reduced costs, simplified and streamlined scale-up, and flexibility compared to batch process alternatives.

This research presents an approach to control system design for the continuous production of pharmaceuticals. First, a dynamic model of the continuous drug manufacturing process is developed that leads to a system of nonlinear, differential algebraic equations. This model is then linearized around typical process operating conditions. The linearized model is subsequently discretized and is used to develop a real time online model predictive control system that ensures the integrity of the pharmaceutical production process and helps satisfy the stringent FDA requirements typically required for safe drug production. Specifically, a constrained infinite time linear quadratic regulator control strategy is employed that can satisfy hard constraints in real time. For example, if there is variability in the quality of raw materials, the controller can ensure, in real time, that FDA requirements on drug purity are met.

The proposed approach is demonstrated with a case study based on a continuous Ibuprofen manufacturing process, as developed for steady-state conditions by Jolliffe & Gerogiorgis (2014), and references therein. In this work, a dynamic model of this process is developed, linearized, and controlled as described above.

534547 Model Predictive Control of a Rotational Molding Process

Abstract:

Rotational Molding, also known as rotomolding or rotational casting, is a batch process used for the processing of plastics through distinct heating and cooling phases [11]. In this process a mold filled with a powdered charge is slowly rotated in a heated oven, resulting in the softened material being dispersed and sticking to the walls of the mold. In order to maintain an even wall thickness, the mold is continuously rotated at all times both during the heating and cooling phase. One of the control challenges is to obtain consistent high-quality products from one batch to another while avoiding unfinished parts with incomplete curing or degradation due to prolonged overheating.

Traditional approach for control of rotational molding processes has been to use an open loop, recipe-based policy. In this approach, a predefined input trajectory is applied to each batch [1]. The approach assumes that the desired product quality can be obtained by repeating historically successful input profiles. Although these approaches are easy to implement and do not require a model for the process, they are incapable of rejecting disturbances that affect the process. This has motivated the use of feedback control strategies.

A classical closed loop control approach that could be used in the rotomolding process is trajectory tracking. In this approach, a predefined set-point trajectory for a measured process variable, such as internal mold air temperature, is tracked. PI controllers are often used for implementing trajectory tracking control. The challenge with trajectory tracking approach is that even with perfect tracking, it may not result in desired quality, as the relationship between the measured/tracked variable and final quality may change significantly with changes in the process conditions across batches.

These challenges are addressed by control strategies that are cognizant of the causal relationship, not only between the manipulated and the (online) measured variables, but also between the manipulated and the final quality variables. A popular model-based control approach, model predictive control (MPC), has increasingly been studied for the control of batch processes [2-5]. The reasons for the popularity of these control schemes are twofold: first, the feedback controller can counter the model uncertainties that are associated with model simplifications and measurement errors; the other being their ability to handle the ever-present input/output constraints. In implementing these formulations, where possible, good first principles models are preferred due to the ability to predict process behavior beyond the data set used and estimate model parameters. However, first principles models are difficult to develop, as it is often impractical to estimate the parameters through experiments, particularly because of the associated cost. Furthermore, these first principles models pose computational challenges for direct use in predictive control formulation, for instance, in case of distributed parameter systems such as batch particulate processes.

These issues have motivated the use of simpler, often linear, models derived from past batch data. A variety of approaches for the development of data-driven models have been proposed. One excellent approach is partial least squares (PLS), which models the process in a projected latent space [6]. These models are essentially time-varying linear models, linearized around mean past trajectories, and therefore require the batches to be the same length, or to recognize an appropriate alignment variable. To account for these limitations, a multi-model approach was proposed in [5]. The multiple models were based on the 'current measurements' of the process instead of the 'time'. These developments were followed by contributions in the area of integration of these data-driven models with the advanced control formulations [4-5]. More recently, a subspace identification based batch control approach was proposed in [7] where a LTI state-space model of the batch process is estimated. Subspace identification methods, in general, are a class of model identification methods that is non-iterative in nature and models a process as a state-space linear time invariant system, identifying the system matrices along with the order of the system. One of the major advantages with this approach is that it facilitates the accommodation of variable duration batches without the need for alignment of batch lengths as demonstrated through different applications in [7-10].

Motivated by these considerations, this work presents a subspace identification based state-space modeling and control approach for rotational molding process to achieve desired quality specifications. Contributions from this developed system represent the progression of this industry towards advanced manufacturing practices, with improved reliability between batches.

[1] D. Bonvin, B. Srinivasan, D. Hunkeler, Control and optimization of batch processes, IEEE Control Systems Magazine 26 (6) (2006) 34-45.

[2] D. Shi, N. H. El-Farra, M. Li, P. Mhaskar, P. D. Christofides, Predictive control of particle size distribution in particulate processes, Chemical Engineering Science 61 (1) (2006) 268-281.

[3] D. Bonvin, Optimal operation of batch reactors- a personal view, Journal of Process Control 8 (5-6) (1998) 355-368.

[4] S. Aumi, P. Mhaskar, Integrating Data-Based Modeling and Nonlinear Control Tools For Batch Process Control, AIChE Journal 58 (2012) 2105-2119.

[5] S. Aumi, B. Corbett, P. Mhaskar, and T. Clarke-Pringle. Data-based modeling and control of nylon-6, 6 batch polymerization. Control Systems Technology, IEEE Transactions on, 21(1):94-106, Jan 2013.

[6] J. Flores-Cerrillo, J. F. MacGregor, Control of particle size distributions in emulsion semibatch polymerization using mid-course correction policies, Industrial and Engineering Chemistry Research 41 (7) (2002) 1805-1814.

[7] B. Corbett, P. Mhaskar, Subspace identification for data-driven modelling and quality control of batch processes, AIChE Journal 62 (2016) 1581-1601.

[8] Abhinav Garg, P. Mhaskar (2017). Subspace Identification Based Modeling and Control of Batch Particulate Processes, Industrial and Engineering Chemistry Research, 56 (26) 7491-7502.

[9] Abhinav Garg, Brandon Corbett, Prashant Mhaskar, Gangshi Hu, Jesus Flores-Cerrillo (2017), High Fidelity Model Development and Subspace Identification of a Hydrogen Plant Startup Dynamics, Computers and Chemical Engineering, 106 183-190.

[10] Rashid, M. M., Mhaskar, P., & Swartz, C. L. (2017). Handling multiâ€rate and missing data in variable duration economic model predictive control of batch processes, AIChE Journal, 63(7), 2705-2718.

[11] Felipe P.C. Gomes, Abhinav Garg, P. Mhaskar and Michael R. Thompson, Quality monitoring of rotational molded parts using a nondestructive technique, ANTEC (2018), Accepted.

534349 A Multiparametric Programming Based Approach to Integrate Design, Scheduling, and Control of a Batch Process

Abstract:

Integrating short term (control), middle term (schedule), and long term (design) decisions has been shown to improve the operability and profitability of batch processes [1]. The interactions between these distinct time-scale decisions allow for opportunities to (i) replace fixed recipe-based schedules with detailed model-based schedules, (ii) build design dependent schedule aware controllers, and (iii) reduce capital costs by circumventing overdesigned processing units. While the efficiency of simultaneous decision making in process systems is greatly acknowledged, the methodology/procedure to achieve an effective integration is still an open question due to the order of magnitude differences in the time scales of the constituent problems, as well as their respective objectives. The most recent advances within the last decade [2-7] depict a promising basis for further improvement.

In this work, we present a unified framework to simultaneously determine the optimal codependent short term and middle term operating strategies of a batch process for a range of market conditions, while minimizing the capital costs. A single high fidelity model is utilized to develop (i) multiparametric rolling horizon optimization (mpRHO) strategies as a function of process states and time variant market conditions, and (ii) multiparametric Model Predictive Control (mpMPC) for effective set-point tracking, following a similar methodology described in Burnak *et al* [8]. The design of the process is accounted for as an additional unknown bounded parameter in the offline operational strategies. The offline nature of mpRHO and mpMPC strategies allows for integration in the design optimization of the process, i.e. minimizing the capital costs. The framework is showcased on (i) a single non-isothermal batch reactor, and (ii) a batch process comprising a preprocessing unit, a reactor, and a post-processing unit.

1. Pistikopoulos, E. N., Diangelakis, N. A., Towards the integration of process design, control and scheduling: Are we getting closer?, Computers and Chemical Engineering (2015), 91, pp. 85-92.
2. Zhuge, J., Ierapetritou, M. G., Integration of scheduling and control for batch processes using multi-parametric model predictive control, AIChE J. (2014), 60 (9), pp. 3169-3183.
3. Baldea, M., Harjunkoski, I., Integrated production scheduling and process control: A systematic review, Computers and Chemical Engineering (2014), 71, pp. 377-390.
4. Dias, L. S., Ierapetritou, M. G., Integration of scheduling and control under uncertainties: Review and challenges, Chemical Engineering Research and Design (2016), 116, pp. 98-113.
5. Patil, B.P., Maia, E., Ricardez-Sandoval, L.A., Integration of scheduling, design, and control of multiproduct chemical processes under uncertainty, AIChE Journal (2015), 61, pp. 2456-2470.
6. Dias, L. S., Pattison, R. C., Tsay, C., Baldea, M., Ierapetritou, M. G., A simulation-based optimization framework for integrating scheduling and model predictive control, and its application to air separation units, Computers and Chemical Engineering (2018), pp. 139-151.
7. Rossi, F., Casas-Orozco, D., Reklaitis, G., Manenti, F., Buzzi-Ferraris, G., A computational framework for integrating campaign scheduling, dynamic optimization and optimal control in multi-unit batch processes, Computers and Chemical Engineering (2017), 107, pp. 184-220.
8. Burnak, B., Katz, J., Diangelakis, N. A., Pistikopoulos, E. N. Simultaneous process Scheduling and control: A multiparametric programming based approach, Industrial & Engineering Chemistry Research (2018), 57 (11), 3963-3976.

524393 Multi-Objective Dynamic Optimisation of a Fed-Batch Nosiheptide Reactor

Abstract:

Thiopeptide antibiotics are highly potent and structurally complex; this growing class of secondary metabolites are formed by ribosomal peptide biosynthesis. Nosiheptide, a sulfur-containing peptide antibiotic obtained through fermentation, can be regarded as the structural prototype of the class e thiopeptide antibiotics. It exerts exceptional antibiotic activity in vitro and in a mouse model against critical Gram-positive pathogens such as MRSA, VRE, or *Clostridium difficile*. Shown non-toxic at high dosages it has previously been used for applications in farm animals (Benazet at al., 1980). Recently the first total synthesis of nosiheptide was reported, utilizing double macrocyclization of a fully functionalized linear precursor (Wojtas et al., 2016). Given low indusial yields, strong motivation exists to dynamically optimise the process for improved product yield while reducing production time and cost. Similar multi-objective optimisation problems are common in the chemical industry (Rodman, Fraga and Gerogiorgis, 2017), which can be solved with both stochastic and deterministic techniques.

The nosiheptide fermentation process is an extremely complex biochemical reaction, with the complete reaction mechanism not strictly understood. A simplified mechanistic fermentation process model has been proposed, considering only the most essential interactions in the biochemical network (Wojtas et al., 2016). The model considers the temperature and pH dependence of biomass growth, in addition to substrate consumption and product production, with the 19 model parameters determined from the model authors experimental campaigns.

To circumvent mass transfer inhibition at excessive substrate concentrations in the broth the nosiheptide reactor is operated in a fed-batch mode, where the reactor is only partially filled initially, with substrate supplemented over time. Herein the multi-objective dynamic optimisation problem can be explicitly formulated: our two minimisation criteria are batch time and inverse yield with four decision (control) profiles to compute towards optimising the system: both the dynamic feed flow rate and the dissolved oxygen concentration must be determined in addition to the reactor temperature and pH profile subject to operability constraints.

This contribution employs (epsilon constrained ***or*** variable weighted sum) handling of the bi-objectives problem, allowing the Pareto front trade-off between the competing objectives to be visualised. A direct method for dynamic optimisation (simultaneous strategy) has been performed in each case to compute the four optimal control trajectories. Orthogonal polynomials on finite elements are used to approximate the control and state trajectories allowing the continuous problem to be converted to NLP form (Biegler, Cervantes, and Wächter, 2007). The resultant large scale NLP problem is been solved for each instance using IPOT (Wächter and Biegler, 2006), and global optimality is ensured with a multi-start search via Latin Hypercube Sampling of the input space for initialisation.

The algorithm employed demonstrates that appropriate reactor operation and control can significantly improve product yield. Additionally, explicit visualization of the objective trade-offs provides valuable insight to decision makers towards critical economic process decisions.

REFERENCES

Benazet, F., Cartier, M., Florent, J., et al., 1980. Nosiheptide, a sulfurcontaining peptide antibiotic isolated from Streptomyces actuosus 40037. *Cell. Mol. Life Sci.* **36**(4): 414â"'416.

Biegler, L. T., Cervantes, A. M. and WÃ¤chter, A., 2002. Advances in simultaneous strategies for dynamic process optimization. *Chemical Engineering Science* **57**: 575-593.

Biegler, L.T., 2007. An overview of simultaneous strategies for dynamic optimization. *Chemical Engineering and Processing: Process Intensification* **46**(11): 1043-1053.

Niu, D., Jia, M., Wang, F. and He, D., 2013. Optimization of nosiheptide fed-batch fermentation process based on hybrid model. *Industrial & Engineering Chemistry Research* **52**(9): 3373-3380.

Rodman, A.D. and Gerogiorgis, D.I, 2017. Dynamic optimization of beer fermentation: sensitivity analysis of attainable performance vs. product flavor constraints. *Computers and Chemical Engineering* **106**(C): 582-595.

Rodman, A.D., Fraga, E.S. and Gerogiorgis, D.I., 2018. On the application of a nature-inspired stochastic evolutionary algorithm to constrained multi-objective beer fermentation optimisation. *Computers and Chemical Engineering* **108**: 448-459.

Wojtas, K.P., Riedrich, M., Lu, J.Y., Winter, P., Winkler, T., Walter, S. and Arndt, H.D., 2016. Total Synthesis of Nosiheptide. *Angewandte Chemie International Edition*, **55**(33): 9772-9776.

534565 Big Data and the Chemical Industry: From Sensor to Value

Abstract:

With the Internet of Things, smart and wireless sensors, advanced control systems, wireless communications and smart devices, the amount of data collected and stored has grown exponentially. These bring big opportunities to the chemical industry where Big Data can be used to enhance manufacturing & operational excellence, forecast pricing, improve formulations, and improve marketing and support innovation.

We present the challenges and opportunities of Big Data in SABIC and a case study illustrating the use of BD technology to move from data, to knowledge, to value in support of Innovation.

536696 Big Data and Iot: A Demonstration Testbed of Multi-Stage Centrifugal Pumping System

Abstract:

Big data and IoT: a demonstration testbed of multi-stage centrifugal pumping system

Devarshi Shah[1], Q. Peter He[1], and Jin Wang[1]

[1]Department of Chemical Engineering, Auburn University, Auburn, AL, 36849, USA

Abstract:

A new industrial revolution is happening to make current manufacturing systems smarter, safer, and more efficient. With the emergence of the Industrial Internet of Things (IoT) and ever advancing computing power and expansion of wireless networking technologies, a new generation of networked, information-based technologies, data analytics, and predictive modeling are providing unprecedented embedded computing capabilities as well as access to previously unimagined potential uses of data and information. This revolution is known as smart manufacturing [1] or Industry 4.0.

Although many general frameworks have been proposed for IoT enabled systems for industrial application [1]-[3]. There is limited literature on demonstrations or testbeds of such systems. In addition, there is a lack of systematic study on the characteristics of IoT sensors and data analytics challenges associated with IoT sensor data. This study is an attempt to help fill this gap by exploring the characteristics of IoT vibration sensors and show how IoT sensors and big data analytics can be used to develop real time monitoring frameworks.

In this work, we will introduce the design of an IoT testbed using multi-stage centrifugal pumping system equipped with non-invasive IoT vibration sensors as shown in Fig. 1. We will discuss the system architecture and protocols designed for data collection, transmission, and storage as shown in Fig. 2. We will also discuss the characteristics of IoT sensors and data analytics challenges associated with data collected from these sensors. The main focus of this work is the development of data-driven predictive models based on the vibration signals to infer flowrate inside the pipe and rpm of the pump motor, and compare with experimental measurements. Altogether, this study serves as a demonstration of how IoT sensors and big data analytics can be integrated and utilized for real-time process monitoring.

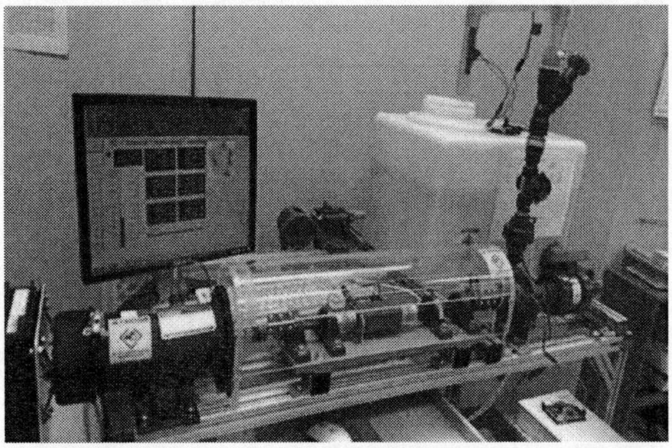

Fig. 1 Actual setup of the testbed with five IoT vibration sensors attached to the pumping system at five different locations

References:

[1] J. Davis, T. Edgar, J. Porter, J. Bernaden, and M. Sarli, "Smart manufacturing, manufacturing intelligence and demand-dynamic performance," *Comput. Chem. Eng.*, vol. 47, pp. 145–156, Dec. 2012.

[2] G. Fortino and P. Trunfio, Eds., *Internet of Things Based on Smart Objects*. Cham: Springer International Publishing, 2014.

[3] H. S. Kang, J. Y. Lee, S. Choi, H. Kim, J. H. Park, J. Y. Son, B. H. Kim, and S. Do Noh, "Smart manufacturing: Past research, present findings, and future directions," *Int. J. Precis. Eng. Manuf. Technol.*, vol. 3, no. 1, pp. 111–128, Jan. 2016.

536078 Pressure Swing Adsorption Cycle Synthesis Utilizing Artificial Neural Networks As Surrogate Models

Abstract:

With fossil fuels expected to be a significant portion of the worldâ€™s energy mix for the near future, it is important to minimize CO_2 emissions from current power plants through carbon capture and sequestration (CCS).[1] In post-combustion CCS, CO_2 is separated from the power plant flue gas emissions, containing mainly of N_2 and CO_2. Of the technologies currently available for CCS, pressure swing adsorption (PSA) is one of the more promising due to low energy requirements and short cycle times compared to other adsorption based technologies. However, one challenge that exists in using PSA for CCS is designing the cycle to match newly developed materials for this application. Although the steps in all PSA cycles can be classified into six different possibilities (pressurization, feed, depressurization, light reflux, heavy reflux and pressure equilibration), the arrangement of the steps and interactions between steps lead to hundreds of potential different combinations. The objective of this work is to develop a new approach that is capable of synthesizing PSA cycles to capture the CO_2 from flue gas at the required purities and recoveries while minimizing energy requirements and maximizing adsorbent productivities.

In this work, we present a new framework for synthesizing the PSA cycle with the lowest predicted CO_2 capture costs. In this framework, we train artificial neural networks (ANNs) using Bayesian regularization methods as surrogate models for the various PSA steps. The ANNs are trained on simulation data collected from our PSA model consisting of a system of partial differential algebraic equations incorporating mass and energy balances, pressure drop across the column, competitive multi-site Langmuir isotherms and the linear driving force model.[2] With the ANN surrogate models, we propose a mixed integer nonlinear programming (MINLP) model to determine the ideal ordering and duration of steps in order to minimize the energy requirements and maximize the adsorbent productivity. We evaluate this model with several adsorbents, including Ni-MOF-74, UTSA-16 and zeolite 13X to compare the adsorbents under optimized cycle conditions.

1. MIT. *The Future of Coal*; 2007. http://web.mit.edu/coal/.
2. Leperi KT, Snurr RQ, You F. Optimization of Two-Stage Pressure/Vacuum Swing Adsorption with Variable Dehydration Level for Postcombustion Carbon Capture. *Ind. Eng. Chem. Res.* 2016;55:3338-3350.

537187 Continuous Control of a Polymerization System with Deep Reinforcement Learning

Abstract:

Improving and deriving noble control laws for polymerization reaction has been one of the uppermost tasks in the chemical industry. There have been extensive studies in control industries from proportional controls to various forms of model predictive controls, which have been benefiting in current chemical processes and plants[1]. Recent breakthroughs in deep learning area have started to inspire the development of Artificial Intelligence(AI)-based controllers in chemical reaction control area thanks to its known success in wide-range applications from gaming to robotics control[2]. These applications utilize deep reinforcement learning (Deep RL), of which the ultimate goal is to enable computers to make human-like policies based on AI's exploration of the environment[3]. However, some challenges are still faced for applying Deep RL in controlling chemical reactions, since the inputs of chemical reactors are usually high-dimensional, and the system dynamics are often sensitive and has considerable time-delay effect[2].

Previous studies have been done in controlling a free-radical polymerization process by following real-time measurement of weight-average molar mass[1], where the specific molar mass distribution can be achieved by following a trajectory of weight-average molar mass with respect to time[1]. In this work, we developed a deep learning based controller for a free radical poly-acrylamide polymerization system using Deep Deterministic Policy Gradient (DDPG). The DDPG utilizes actor-critic structure that is able to predict actions of infinite dimensions[4]. The controller calculates the control action a_t, which consists of the monomer flow rate F_m and initiator flow rate F_i to adjust at each time step t, and the system response of the action is recorded as a state s_t. The network is trained to maximize the cumulative reward r that accounts for the distance between current output and target output for each iteration[5]. The network is trained on an established kinetic model of the polymerization reaction, and the controller gradually learns the policy through exploration of the system. Then convergence is achieved when the average cumulative reward reaches a desired threshold. In our experiment, the controller successfully has learned the control policy to follow the target trajectory of the weight-average molar mass.

Overall the smart controller has shown robust control over a range of operating conditions, which indicates the deep reinforcement learning based approach's capability in controlling a nonlinear dynamic semi-batch system.

Reference

[1] N. Ghadipasha, W. Zhu, J. A. Romagnoli, T. Mcafee, T. Zekoski, and W. F. Reed, "Online Optimal

Feedback Control of Polymerization Reactors : Application to Polymerization of Acrylamide – Water – Potassium Persulfate (KPS) System," 2017.

[2] S. P. K. Spielberg, R. B. Gopaluni, and P. D. Loewen, "Deep Reinforcement Learning Approaches for Process Control", 2017

[3] V. Mnih, D. Silver, and M. Riedmiller, "Playing Atari with Deep Reinforcement Learning," pp. 1–9.

[4] T. P. Lillicrap *et al.*, "Continuous learning control with deep reinforcement learning," , 2016.

[5] D. Silver, G. Lever, D. Technologies, G. U. Y. Lever, and U. C. L. Ac, "Deterministic Policy Gradient Algorithms."

529266 The Age of Application in Bioprinting

Abstract:

Ricky Solorzano is the CEO of Allevi, a Philadelphia-based startup that is revolutionizing biological research by building tools to make it easier to engineer biology. Allevi is specifically beginning by creating premier bioprinters that can print living tissues.

Ricky continues to lead the development of next generation products that allows scientist to design and prototype tissues. Ricky is a member of the Organ Bioengineering Scientific Direction Team for the International Space Station as well as a member on the Advanced Regenerative Manufacturing Institute.

Ricky will speak about some state of the art topics in creating tissues through bioprinting and what the future holds for biomanufacturing. In an era where bioprinting continues to hold promise sometimes its hard to understand why and how are they useful. What key applications will allow me to take my research to the next level and stay on the cutting edge. Come and listen to the key ways bioprinting is being most commonly used by researchers around the world.

520721 Nanoengineered Ionic-Covalent Entanglement (NICE) Bioinks for 3D Bioprinting

Abstract:

Three-dimensional (3D) bioprinting is emerging as a promising method for rapid fabrication of biomimetic cell-laden constructs for tissue engineering using cell-containing hydrogels, called bioinks, that can be crosslinked to form a hydrated matrix for encapsulated cells. However, extrusion based 3D bioprinting has hit a bottleneck in progress due to the lack of available bioinks with high printability, mechanical strength, and biocompatibility. Here, we introduce an enhanced nanoengineered ionic covalent entanglement (NICE) bioink for the fabrication of mechanically stiff and elastomeric 3D biostructures. NICE bioink formulations combine nanocomposite and ionic covalent entanglement (ICE) strengthening mechanisms to print customizable cell-laden constructs for tissue engineering with high structural fidelity and mechanical stiffness. Nanocomposite and ICE strengthening mechanisms complement each other through their synergistic interactions, improving mechanical strength, elasticity, toughness, and flow properties beyond the sum of the effects of both reinforcement techniques alone. Herschel-Bulkley flow behavior shields encapsulated cells from excessive shear stresses during extrusion. The encapsulated cells readily proliferate and maintain high cell viability over 120 days within the 3D-printed structure, which is vital for long-term tissue regeneration. The unique aspect of the NICE bioink is its ability to print much taller and higher aspect ratio structures than conventional bioinks without requiring secondary supports. We envision that NICE bioinks can be used to bioprint complex, large-scale, cell-laden constructs for tissue engineering with high structural fidelity and mechanical stiffness for applications in custom bioprinted scaffolds and tissue engineered implants.

537406 Engineering a Highly Elastic Protein-Based Bioink for Printing Complex Soft Tissues

Abstract:

Introduction

Bioprinting has shown remarkable promise for fabrication of complex tissues and organs, and has tremendous potential to overcome issues with organ transplant shortage, drug testing and screening, and the study of biological phenomena such as tissue morphogenesis [1]. The most common hydrogel-based bioinks are biodegradable hydrogels, including those made of naturally derived polymers, such as fibrin, gelatin, hyaluronic acid, alginate, and agarose; synthetic polymers, such as poly(ethylene glycol) diacrylate (PEGDA) [2]; or natural–synthetic composites [3-7]. However, many of these hydrogels are brittle and lack elasticity due to the nature of the polymer backbones, which limits their ability to mimic the mechanical behavior (e.g. softness, stretchability, and elasticity) of human tissues, such as skin, lung, skeletal muscle, and cardiovascular tissues. Elastin-based proteins have been shown to exhibit enhanced elasticity as compared to many other polymers, as well as high tunability of the polymer backbone using recombinant technologies. In this study, we developed a highly elastic bioink for 3D printing by combining a photocrosslinkable elastin-based protein and silicate nanoparticles (Laponite XLG). This novel bioink can be used for fabrication of complex elastic tissues with exquisitely defined architectures, such as lung, muscle, vascular, and cardiac tissues.

Materials and methods

The elastic protein-based bioink was synthesized using a composite composed of elastin-based proteins obtained via recombinant technology [8, 9] and Laponite, which impart liquid polymer solutions with enhanced shear thinning properties. The polymer solution and nanoparticles were mixed in a solution containing lithium phenyl-2,4,6-trimethylbenzoylphosphinate (LAP) (0.5% w/v) in distilled water at 4 °C. The resulting solution was kept at 4 °C prior to bioprinting to prevent coacervation and aggregation of the elastin-based protein. For experiments with cells, a concentrated solution of cells was added to the bioink using a static mixing device (CELLINK). The mechanical and rheological properties of the composite bioinks were characterized based on procedures described elsewhere [10, 11]. After careful mixing, the bioink was quickly transferred to a 3 mL syringe cartridge and stored in a custom cooled printhead maintained at 4 °C during the printing process. A 27 Gauge printing nozzle (210 µm inner diameter) was fixed onto the cartridge before printing. Printing pressures and speeds were adjusted depending on ink flow properties, and gels were typically printed at pressures ranging from 40-100 kPa at printing speeds between 1-10 mm/s. Different geometric shapes were printed by extruding the elastin protein/Laponite solution through a 27 Gauge nozzle and then photopolymerized for 120 sec under exposure to visible light (395–405 nm, 10 mW/cm^2).

Result and discussion:

The mechanical and rheological properties of hydrogels with varied concentrations of elastin protein and Laponite were evaluated prior to printing. The optimal composition was chosen for its high elasticity and shear-thinning properties. Various structures relevant to elastic human tissues were printed, including a blood vessel, a left ventricle, a bronchial tree, and a simple lattice structure that mimics the aligned syncytium of muscle cells. The resolution could be adjusted by varying the printing pressure or printing speed, where increased speed and decreased pressure provided smaller printed lines. Cyclic mechanical testing of printed structures demonstrated that the constructs exhibited high recovery of their original shape with low energy loss and hysteresis. We printed hydrogels loaded with cells to determine the practicality of using this material for bioprinting cell–laden constructs. High cell viability was achieved in the printed constructs over the course of a week. These results suggest that the elastin-based bioink could potentially be used to print a wide range of different human elastic tissues.

Conclusion

In this study, we developed a visible light crosslinkable and highly elastic bioink based on an elastin-based protein and Laponite. To our knowledge, this is the first time that elastin-based composites have been used to bioprint complex tissue-like structures. The engineered bioprinted constructs showed high biocompatibility, slow degradability, and easily tunable mechanical properties. In addition, the resilience of the engineered constructs can provide a novel material platform for soft tissue engineering, especially elastic tissues such as lung, arteries and heart.

Acknowledgements

The authors would like to acknowledge the support from the American Heart Association (AHA, 16SDG31280010) and National Institutes of Health (R01- EB023052; R01HL140618).

References

[1] S.V. Murphy, A. Atala, 3D bioprinting of tissues and organs, Nat. Biotechnol. 32(8) (2014) 773-785.

[2] T. Jungst, W. Smolan, K. Schacht, T. Scheibel, J. Groll, Strategies and Molecular Design Criteria for 3D Printable Hydrogels, Chem. Rev. 116(3) (2016) 1496-1539.

[3] C. Xu, W. Lee, G. Dai, Y. Hong, A Highly Elastic Biodegradable Single-Network Hydrogel for Cell Printing, ACS applied materials & interfaces (2018).

[4] X. Cui, T. Boland, Human microvasculature fabrication using thermal inkjet printing technology, Biomaterials 30(31) (2009) 6221-6227.

[5] X. Wang, Y. Yan, Y. Pan, Z. Xiong, H. Liu, J. Cheng, F. Liu, F. Lin, R. Wu, R. Zhang, Generation of three-dimensional hepatocyte/gelatin structures with rapid prototyping system, Tissue engineering 12(1) (2006) 83-90.

[6] V. Lee, G. Singh, J.P. Trasatti, C. Bjornsson, X. Xu, T.N. Tran, S.-S. Yoo, G. Dai, P. Karande, Design and fabrication of human skin by three-dimensional bioprinting, Tissue Engineering Part C: Methods 20(6) (2013) 473-484.

[7] N.E. Fedorovich, J.R. De Wijn, A.J. Verbout, J. Alblas, W.J. Dhert, Three-dimensional fiber deposition of cell-laden, viable, patterned constructs for bone tissue printing, Tissue Engineering Part A 14(1) (2008) 127-133.

[8] R.P.-L. Ehsan Shirzaei Sani, Andrew Spencer, Wendy Yu, Benjamin M. Geilich, Iman Noshadi, Thomas J. Webster, Nasim Annabi, Engineering adhesive and antimicrobial hyaluronic acid/elastin like polypeptide hybrid hydrogels for tissue engineering applications, ACS Biomaterials Science & Engineering (in press).

[9] N. Annabi, S.R. Shin, A. Tamayol, M. Miscuglio, M.A. Bakooshli, A. Assmann, P. Mostafalu, J.Y. Sun, S.

Mithieux, L. Cheung, X. Tang, A.S. Weiss, A. Khademhosseini, Highly Elastic and Conductive Human-Based Protein Hybrid Hydrogels, Adv. Mater. 28(1) (2016) 40-49.

[10] N. Annabi, D. Rana, E.S. Sani, R. Portillo-Lara, J.L. Gifford, M.M. Fares, S.M. Mithieux, A.S. Weiss, Engineering a sprayable and elastic hydrogel adhesive with antimicrobial properties for wound healing, Biomaterials 139 (2017) 229-243.

[11] S. Hong, D. Sycks, H.F.a. Chan, S. Lin, G.P. Lopez, F. Guilak, K.W. Leong, X. Zhao, 3D Printing: 3D Printing of Highly Stretchable and Tough Hydrogels into Complex, Cellularized Structures, Advanced materials (Deerfield Beach, Fla.) 27(27) (2015) 4034-4034.

533636 Silk Protein-Based Hydrogels for 3D Printing of Tissue Constructs

Abstract:

Substrate topography, along with mechanical properties and presence of cell-relevant ligands and cues, has been established as a crucial design parameter that affects cellular behavior in *in vitro* tissue models. Understanding cellular behavior in engineered human tissue is expected to lead to greater physiological relevance of these models, which may ultimately facilitate application of the model for *in vitro* drug screening or disease modeling. In the small and large intestine, villi and crypts, respectively, are the major morphological features that contribute to tissue function. Recent examples of *in vitro* tissue models have sought to incorporate crypt topographies more representative of large intestine morphology into culture substrates through different means, including the use of decellularized extracellular matrix, formation of hydrogels around sacrificial components, or casting scaffolds against patterned substrates. While previous work with these models have shown that topography affects cellular response, there are two main limitations with these models: 1) the constructs degrade quickly in many cases, and 2) the topography cannot be easily varied. This research focuses on preparing intestine-like structures by 3D printing silk protein hydrogels. Silk proteins can be enzymatically crosslinked into biocompatible hydrogels with long-term stability and tunable mechanical properties. This work develops a new synthetic approach that enables 3D printing of the enzymatically-crosslinked hydrogels. Rheological experiments were used to determine flow behavior and gelation kinetics, and constructs were then 3D printed using robotic dispensing. Intestinal cell lines were cultured on and within the gels to quantify cell attachment, proliferation, and morphology on the printed constructs.

522421 Bioprinting of Large-Scale Hydrogels with Build-in Vascular Channels

Abstract:

Creating three dimensional (3D) vascular networks within tissue-engineered scaffolds is one of the major challenges in tissue engineering and regenerative medicine. Although there are some successful examples of vascularized channels, there is still a demand for robust fabrication of high-resolution 3D channel structures within thick (human-scale) scaffolds. 3D bioprinting enables fabrication of 3D scaffolds from multi-materials (bioinks) including hydrogels and cells with spatial resolution. In this work, we used extrusion-based bioprinting to fabricate thick scaffolds with highly tunable channels. Unlike freeform printing approaches, where a hydrogel is printed within another hydrogel or viscous medium, we developed a novel sequential printing approach. In our approach, a sacrificial ink formulation is printed sequentially with a photocurable â€œmatrixâ€ bioink formulation. Sacrificial bioink is printed at the interface of the matrix bioink. UV curing is followed by printing of additional matrix bioink, and final curing. The sacrificial hydrogel is then removed to achieve perfusable, high resolution, freeform channels. This approach allows us to fabricate human-scale scaffolds with tunable channels. In this study, matrix bioinks are formulated from photocurable hyaluronic acid-based or alginate-based hydrogels, with or without human mesenchymal stem cells (hMSC). Sacrificial hydrogel ink is formulated from Pluronic F-127. Human umbilical vein endothelial cells (HUVEC) are seeded within the channel surfaces. Within 7 days, uniform confluent endothelial layers are formed with the help of perfusion leading to a non-leaky layer within the channels. Meanwhile, the matrix embedded hMSCs show high viability during the whole culturing process. Our novel approach shows great potential to rapidly fabricate large scale vascularized scaffolds that can be used as in vitro models or tissue engineered implants.

523591 3D Bio-Printed Model of Brain Tumor Microenvironment with Vasculatures

Abstract:

Introduction:

Glioblastoma multiform (GBM) is a malignant brain tumor. GBMs have a heterogeneous population with cancer stem cells. The GBM stem cells often develop therapeutic resistance against chemotherapy or radiotherapy, followed by tumor recurrences. The tumor cells are also highly invasive and frequently exploit microvessels as guides for migration. These hallmark behaviors of GBM make it even harder to completely cure GBM, as residual cells from the surgical resection of tumor core aggressively invade and cause tumor recurrences.

Tumor microenvironment consisting of various extracellular matrices, vasculatures, mural cells, chemical/mechanical stimulations, etc., plays a crucial role in tumor growth. Proper microenvironments are required to maintain the GBM stemness and to better mimic tumor cell behaviors in the human body. Current gold standard model for GBM studies is *in vivo* animal model, which has limitations on reproducibility, lethality, and discrepancy with human physiology. Current *in vitro* models such as 2D monolayer culture, transwell culture, 3D hydrogel culture, and spheroid/organoid models provide more reproducible and confined microenvironments. However, many models are lack of sufficient area for tumor invasion/migration, thus have limitations on investigating GBM invasive behaviors under various conditions/periods. The lack of functional vasculatures in current *in vitro* models also restricts studying perivascular invasion/migration of GBM.

In this study, we created various types of brain tumor-vascular microenvironments using patient-derived GBM cells with stemness (GSC: glioblastoma stem cells), vasculature endothelial cells (ECs), supporting cells, and hydrogel. 3D bioprinting technology has been utilized to create complex tissues with multiple cell types and matrices with perfusable vasculatures. Three different 3D *in vitro* models developed in the study maintained GBM stemness and recapitulated cell type-specific tumor invasion patterns and drug responses. The results show the potential of the models in studying cellular/molecular interactions between GBM and tumor microenvironment.

Material and Methods:

Bioprinted models were fabricated as previously described [1, 2], using gelatin as a sacrificial material and collagen/fibrin as main scaffold materials. Three different types of GBM cells were tested in the model: SD02 and SD03 are patient-derived GSCs; U87MG is commercially available GBM cell line. For capillary formation (microvessel formation), endothelial cells (human umbilical vein endothelial cells (HUVECs) or brain microvascular endothelial cells) and supporting cells (fibroblasts, pericytes, astrocytes) were mixed within fibrin hydrogel. The cells in 3D environment spontaneously create a capillary network through cell self-assembly. To show the drug test application, Model A was cultured for 26 days until GBM cells invaded into surrounding matrices over 1 millimeter, then 10-100μM of Temozolomide was injected through the large vessels in Model A for next 31 days.

Results and Discussion:

Develop Three Brain Tumor-Vascular Niche Models

Model A consists of large vasculatures and a GBM spheroid (Figure 1A). This model is designed to investigate the interactions between GBM and extracellular matrices and to measure the diffusion pattern from the large fluidic vessels. Model B consists of capillary network and GBM spheroid (Figure 1B) to mimic microenvironment of GBM tumor core. The microvessels in the model can provide a more physiological environment as well as invasion/migration trails for GBM cells. Model C consists of capillary network and scattering GBM single cells (Figure 1C). The model is designed to mimic tumor microenvironments after surgical resection of tumor core. Since GBM cells are highly invasive, the residual tumor cells left from the surgical resection often invade, proliferate, and cause recurrences. The three models had been successfully fabricated, culture over 3-8 weeks, and utilized for various applications.

Model A: Drug Treatment Application

SD02 spheroid, cultured in the setting of Model A, aggressively invaded into the surrounding matrices for 26 days of pre-treatment culture (Figure 1 A*ii*-A*iii*). During the first 3-4 weeks of Temozolomide drug treatment, the invaded cells were gradually regressed and the tumor core was shrinking (Figure 1 A*iv*). However, some SD02 cells survived from the treatment and resumed aggressive invasion even with the continuing treatment (Figure 1 A*v*). The long-term drug treatment of Model A recapitulates the development of therapeutic resistance under chemotherapy.

Model B&C: Different Tumor Invasion Patterns & Cell-Cell Interactions

In the Models B&C, three GBM cell types presented different patterns of invasions and vascular interactions (Figure 1B-C). SD02 cells showed spiky, spreading branch-like invasions (Figure 1 B*ii*). The cells also restricted capillary formation in both spheroid and single cell form (Figure 1 B*ii* & C*ii*). SD03 spheroid has a more dispersed invasion pattern than SD02 and also presented a large number of single cell migrations (Figure 1 B*iii*). SD03 cells showed the most perivascular interactions. The cells spread along with capillaries (Figure 1 B*iii* & C*iii*). Some tumor cells formed a cluster near the capillary branching area (Figure 1 C*iii*). SD03 tumor spheroid did not restrict the capillary formation (Figure 1 B*iii* & C*iii*). The non-stem GBM cell line U87MG showed a lumpy invasion pattern and caused the death of neighboring ECs in Model B (Figure 1 B*iv*). But in Model C, U87MG cells presented a scattered pattern of proliferation and wrapped around the capillaries without inducing EC death (Figure 1 C*iv*).

In Model B, the tumor invasion patterns of all three cell types are similar to that of *in vivo* mouse models with injected GBM cells (data not shown). Also, the patient-derived GSCs culture in Model B maintained its stemness over three weeks of the culture period. Putting the cell type-specific invasions from Model B and the long-term drug responses demonstrated in Model A, our models has a high potential in establishing a personalized therapeutic strategy.

While Model B is closely mimicking *in vivo* models, Model A and Model C provides unique culture environments that cannot be created in current *in vivo* models. The models have a promise for designing and building customized microenvironments for cellular/molecular interaction studies which are hard to be performed on *in vivo* models.

By using the three models presented in this study, we have generated various combinations of culture conditions with following parameters: 1) with or without microvessels; 2) tumor spheroid vs. single cells; 3) GBM cells with or without stemness; and 4) with or without drug perfusion. In addition to these factors, we can modify the scaffold hydrogel compositions, vascular cell types, surrounding cell types, fluidic perfusion through the vasculatures, etc. Thus, our models have a great expandability in diverse *in vitro* research fields and can serve as a valuable tool to model various co-culture conditions and to efficiently control microenvironment.

Conclusions:

This study demonstrated the development and several applications of 3D brain tumor microenvironment with vascultures, shifting the research focus to an essential area of tumor microenvironments. The GBM-vascular models responded differently to GBM cell types, showing its potential in investigating patient-specific tumor behaviors under chemo-/radio-therapy conditions and consequentially helping to tailor patient-specific treatment strategy. In addition, the 3D models can be adapted to other biological systems and serve as a valuable tool to fabricate customized microenvironments.

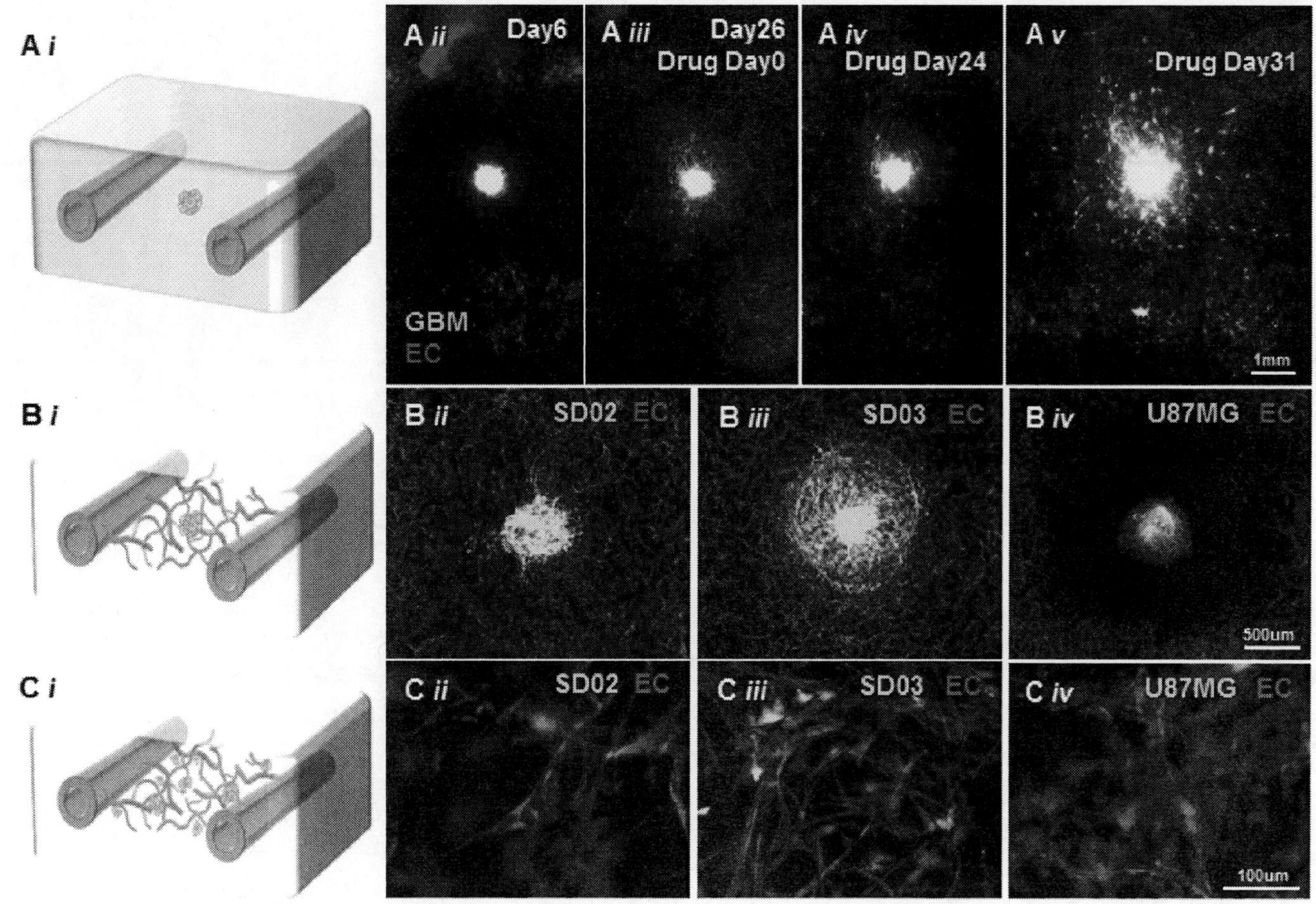

Figure 1. **(A)** Model A with a tumor spheroid and large vasculatures. (A*i*) Schematics. (A*ii*-A*v*) GBM response to Temozolomide treatment. (A*ii*-A*iii*) GBM cells invaded into the surrounding matrices during 26 days of pre-treatment phase. (A*iv*) GBM tumor core was regressed during the first three weeks of drug treatment. (A*v*) The tumor cells developed therapeutic resistance after long-term drug treatment then aggressively invaded into the surrounding tissue area. **(B)** Model B with a tumor spheroid and capillaries. (B*i*) Schematics. (B*ii*) Patient-derived SD02 cells presented spiky, spreading branch-like invasion patterns with restricted capillary formation around the tumor spheroid. (B*iii*) Patient-derived SD03 cells presented more dispersed invasion pattern with a lot of single cell migrations. SD03 tumor spheroid did not restrict the capillary formation. (B*iv*) Commercially available GBM cell line U87MG showed lumpy invasion pattern. The cells often caused endothelial cell deaths around the tumor spheroid. **(C)** Model C with dispersed tumor cells and capillaries. (C*i*) Schematics. (C*ii*) SD02 cells restricted capillary formation and did not present many cell-cell interactions with endothelial cells. (C*iii*) SD03 cells spread along with capillaries. Some tumor cells formed a cluster near the capillary branching area. (C*iv*) U87MG cells presented a scattered pattern of proliferation and wrapped around the capillaries without inducing EC death.

References:

[1] Lee, V. K., Kim, D. Y., Ngo, H., Lee, Y., Seo, L., Yoo, S. S., ... & Dai, G. (2014). Creating perfused functional vascular channels using 3D bio-printing technology. Biomaterials, 35(28), 8092-8102.

[2] Lee, V. K., Lanzi, A. M., Ngo, H., Yoo, S. S., Vincent, P. A., & Dai, G. (2014). Generation of multi-scale vascular network system within 3D hydrogel using 3D bio-printing technology. Cellular and molecular bioengineering, 7(3), 460-472.

537113 A 3D Printed Microfluidic Bioreactor to Engineer Biphasic Construct

Abstract:

INTRODUCTION: Tissue engineered constructs coupled with high-throughput bioreactors can be used as innovative *in vitro* pre-clinical models of tissue functions and disease pathogenesis for drug screening and toxicity assessment [1]. Such *in vitro* models are ideally suited to study tissue-tissue interactions in health and disease in composite tissue units such as those found in the musculoskeletal system. Among these, the osteochondral junction is of particular clinical importance as it has been identified as a locus of degenerative joint diseases such as osteoarthritis [2]. To generate *in vitro* model of adjacent tissues, we have developed an innovative 3D printed bioreactor for biphasic constructs, which can host cells in a small volume and allow optical access for direct monitoring of cellular responses during culture and differentiation in 3 dimensions [3]. The aim of this work is to obtain an osteochondral junction model from adult human mesenchymal stem cells (hMSCs), after delivery of spatially controlled local stimuli.

METHODS: A combined modelling/experimental approach was employed to design a viable bioreactor fluidics. Models of the flow path were created using SolidWorks and tested using ANSYS Fluid Flow (CFX), with an inlet volume flow rate of 1 ml/day, and outlet open to the environment. The central chamber that will host the constructs was considered as filled with methacrylated gelatin (gelMA) with a permeability of $1*10^{-16}$ m^2 and a porosity of 0.8 [4]. Velocities were measured in CFX Post and the design was altered by changing the diameter and height of the channel to achieve flow rates in the constructs not damaging for cells. Bioreactors were designed using SolidWorks and printed with a Viper SLA System at high definition with a layer resolution of 50µm using the Somos® WaterClear Ultra 10122 resin (DSM N.V., The Netherlands). O-rings (McMaster-Carr), glass coverslips (Deckglaser), and adhesive films (Professional Label) were used to complete the system. Bone marrow derived hMSCs were seeded in the bioreactors chambers in gelMA (10^6 cell/ml) and photocrosslinked *in situ*. The hydraulic circuit hosted the incubator, consists of a pump, a 10ml syringe filled with expansion medium, a valve to remove air bubbles, tubes and a collection bag. A dynamic Live Assays was performed to assess cellular viability over time, adding calcein-AM to the medium in the syringe without compromising the overall setup sterility. An Alamar Blue test was also performed within the bioreactor to confirm cell viability. Finally, osteogenic (DMEM, 10% FBS, 5% PSF, 10 nM dexamethasone, 0.1 mM L-ascorbic acid 2-phosphate, and 10 mM beta-glycerophosphate) and chondrogenic (DMEM, 5% PSF, 10 ng/ml recombinant human TGF-β3 (PeproTech), 1% ITS+, 50 µM L-ascorbic acid 2-phosphate, 55 µM sodium pyruvate, and 23 µM L-proline) media were fluxed in the bioreactor for 21 days to induce differentiation which was assessed by histology (Alcian Blue for cartilage, Alizarin Red for bone). Finally, $5x10^3$ hMSCs were plated in a 96 well-plate and incubated with a lentiviral Col2 reporter added at an MOI = 40 (multiplicity of infection = transducing units / #cell) in fresh growth medium without PSF and containing 6 ug/mL of SureEntry transduction reagent. 20h later, medium containing the lentiviral particles was removed and substituted with fresh complete growth medium. Cells were harvested after 72h and added to the non-transfected hMSCs to act as sentinel cells to monitor differentiation.

RESULTS SECTION: Two single-inlet bioreactor designs were first developed with higher and lower perfusion velocities in the chamber hosting the constructs (Fig. 1). The dynamic live assays in the bioreactors produced a successful staining of the cells that can be monitored optically without the need to break sterility in the bioreactor system (Fig. 2). The dynamic Alamar blue assay was also effective in confirming the possibility of non-destructive monitoring within the system. The bioreactor with higher perfusion velocities appeared to be the most effective with the gelMA constructs and that design was adopted for the two inlets, biphasic model. Experimental fluidic tests with the two inlets/outlets biphasic bioreactor confirmed modelling predictions, showing minimal to no mixing between the flowing media (Fig. 3). Differentiation of hMSCs towards a chondrogenic and osteogenic phenotype in the two sides of the constructs was confirmed by histological staining and optically by fluorescence of the Col2 transfected sentinel cells in the half of the construct exposed to chondrogenic medium (data not shown).

DISCUSSION: Development of bioreactors is essential for the generation of biphasic constructs that require specific media for each tissue component and cannot be cultured in regular multiwell plates. The successful dynamic staining of the cells in the constructs within the bioreactors show the potential of the system to perform a continuous, real-time monitoring of constructs maintaining sterility. This is particularly relevant when using sentinel cells with reporter genes that can signal differentiation as well as response to stress and cross-talk between tissues, depending on the specific reporters adopted.

SIGNIFICANCE: Disease modeling and drug screening must account for interactions between adjacent tissues. Once scaled for high throughput, our biphasic bioreactor will be an essential step in developing new strategies to prevent and treat joint diseases.

REFERENCES: [1] Lozito et al. (2013). Stem Cell Research & Therapy, 4(Suppl 1):S6. [2] Alexander et al. (2014). Exp Biol Med, 239(9):1080-95. [3] Nichols et al. (2018) Biomed Microdev 20: 18. [4] Iannetti et al. (2016). PLoS One, 11(9):e0162774.

ACKNOWLEDGEMENTS: Ri.MED Foundation, CASIS (grant no. GA-2016-236), Commonwealth of Pennsylvania, US EPA – Science to Achieve Results (STAR) program #835-73601, NIH (1UG3 TR002136-01)

Fig. 1. Schematics of (A) the bioreactor model with the 2 fluidic channels designs and (B) the two input/output design. Simulation for the lower (C) and higher (D) flow rate designs.

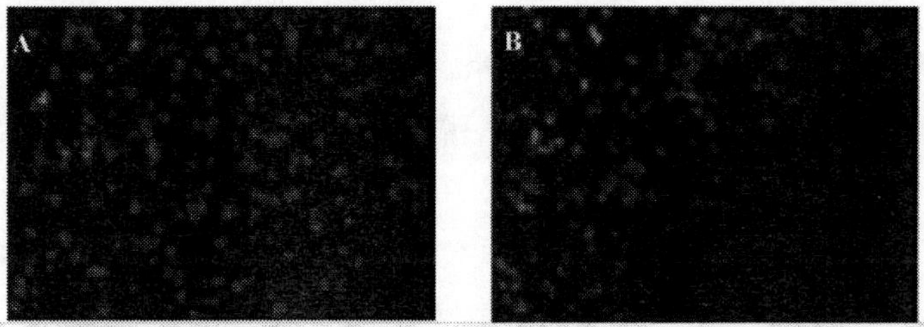

Fig. 2. hMSCs after 1 week in culture in gelMA (A) in the bioreactor and (B) in a petri dish.

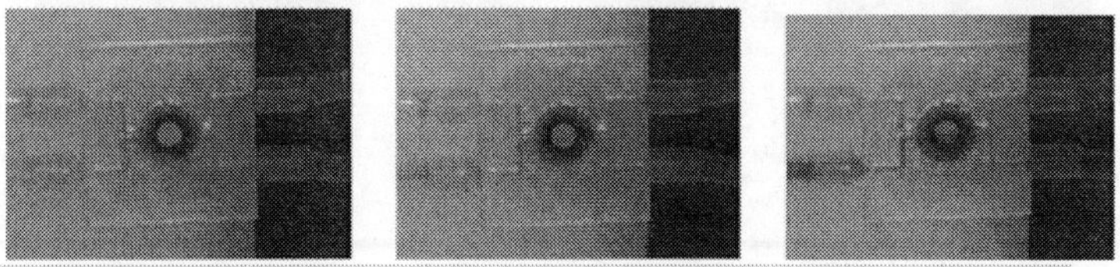

Fig. 3. Time lapse of low within the biphasic bioreactor. No mixing of media occurs.

537785 Photo-Crosslinked Chondroitin Sulfate a and Chitosan for Extruded Vascularization

Abstract:

In tissue engineering, vascularization is a challenge that is mitigated by induced vascularization or artificial vascular structures or a combination of these methods. In this study, we experimented a method to make vascular structures using Chondroitin Sulfate A (CSA), Chitosan and Collagen. We propose this method to be adapted to a 3D bioprinting setup.

CSA and Chitosan make a polyelectrolytic complex that is known to be effective biomaterial for tissue engineered structures. However, the viscoelastic properties of CSA do not allow for good extrusion. In this study, we photo-crosslinked CSA with each other to make a hydrogel with better viscoelastic properties for extrusion. We methacrylated CSA by reacting with methyl anhyrdride and then photo crosslinked under UV. We found that 3% of CSA with a 0.45 degree of methacrylation showed good viscoelastic properties.

This cross-linked CSA hydrogel was supplemented with endothelial cells and extruded into a Chitosan solution to make hollow fibers. This hydrogel was optimized to be extruded into a Chitosan bath of encapsulated cells. The Chitosan was complemented with Collagen to make the membrane of the hollow fibers more susceptible for endothelial cell attachment. This was evaluated with staining and SEM studies. We further evaluated the viability of the method by making the hollow fibers among Mesenchymal Stem Cells that were encapsulated in capsules of CSA-Chitosan complex. We observed the growth and reported the findings.

AUTHOR INDEX

AUTHOR INDEX

AUTHOR INDEX